Laboratory Manual for

CHM 1020

Wright State University
Department of Chemistry

ELEMENTARY ORGANIC CHEMISTRY

Assembled from contributions of the Faculty of the Department of Chemistry.

EDITORS
Travis B. Clark
Kirby A. Underwood

AF174184

Course instructor/contact information: _______________________

Course TA(s)/contact information: _______________________

Fire extinguisher location: _______________________

Shower and eye wash location: _______________________

Nearest telephone location: _______________________

Telephone number for aid: _______________________

Fire escape location from laboratory: _______________________

Nearest AED location: _______________________

VAN-GRINER
LEARNING

Laboratory Manual for

CHM 1020
Elementary Organic Chemistry

Fourth Edition
Editors: Travis B. Clark and Kirby A. Underwood
Wright State University
Department of Chemistry

Printed in the United States of America
10 9 8 7 6 5 4 3
ISBN: 978-1-61740-735-2

Van-Griner Publishing
Cincinnati, Ohio
www.van-griner.com

President: Dreis Van Landuyt
Project Manager: Brenda Schwieterman
Customer Care Lead: Lauren Houseworth

Clark 735-2 Su19
312851-321637
Copyright © 2020

Acknowledgments

The editors are indebted to the writers of all previously published laboratory manuals, the specific contributions of whom are no longer traceable.

Special thanks are due to the members of the Chemistry Department of Wright State University for their comments and suggestions during the preparation of this manual.

We would also like to thank numerous laboratory assistants, who provided their own comments and suggestions.

Table of Contents

Safety Guidelines
1020 Laboratories

WHAT ARE THE SAFETY DOs AND DON'Ts FOR STUDENTS?

Life threatening injuries can happen in the laboratory. For that reason, students need to be informed of the correct way to act and things to do in the laboratory. The following is a safety checklist to acquaint students with the safety dos and don'ts in the laboratory.

Conduct

- Do not engage in practical jokes or boisterous conduct in the laboratory.
- Never run in the laboratory.
- The use of personal audio or video equipment is prohibited in the laboratory.
- The performance of unauthorized experiments is strictly forbidden.
- Do not sit on laboratory benches.

General Work Procedure

- Know emergency procedures.
- Never work in the laboratory without the supervision of a TA.
- Always perform the experiments or work precisely as directed by the TA.
- Immediately report any spills, accidents, or injuries to a TA.
- Never leave experiments while in progress.
- Never attempt to catch a falling object.

- Be careful when handling hot glassware and apparatus in the laboratory. Hot glassware looks just like cold glassware.
- Never point the open end of a test tube containing a substance at yourself or others.
- Never fill a pipette using mouth suction. Always use a pipetting device.
- Make sure no flammable solvents are in the surrounding area when lighting a flame.
- Do not leave lit Bunsen burners unattended.
- Turn off all heating apparatus, gas valves, and water faucets when not in use.
- Do not remove any equipment or chemicals from the laboratory.
- Coats, bags, and other personal items must be stored in designated areas, not on the bench tops or in the aisle ways.
- Notify your instructor or TA of any sensitivities that you may have to particular chemicals if known.
- Keep the floor clear of all objects (e.g., ice, small objects, spilled liquids).

Housekeeping

- Keep work area neat and free of any unnecessary objects.
- Thoroughly clean your laboratory work space at the end of the laboratory session.
- Do not block the sink drains with debris.
- Never block access to exits or emergency equipment.
- Inspect all equipment for damage (cracks, defects, etc.) prior to use; do not use damaged equipment.
- Never pour chemical waste into the sink drains or wastebaskets.
- Place chemical waste in appropriately labeled waste containers.
- Properly dispose of broken glassware and other sharp objects (e.g., syringe needles) immediately in designated containers.
- Properly dispose of weigh boats, gloves, filter paper, and paper towels in the laboratory.

Apparel in the Laboratory

- Always wear appropriate eye protection (i.e., chemical splash goggles) in the laboratory.
- Wear disposable gloves, as provided in the laboratory, when handling hazardous materials. Remove the gloves before exiting the laboratory.
- Wear a full-length, long-sleeved laboratory coat or chemical-resistant apron.
- Wear shoes that adequately cover the whole foot; low-heeled shoes with non-slip soles are preferable. Do not wear sandals, open-toed shoes, open-backed shoes, or high-heeled shoes in the laboratory.

- Avoid wearing shirts exposing the torso, shorts, or short skirts; long pants that completely cover the legs are preferable.
- Secure long hair and loose clothing (especially loose long sleeves, neck ties, or scarves).
- Remove jewelry (especially dangling jewelry).
- Synthetic finger nails are not recommended in the laboratory; they are made of extremely flammable polymers which can burn to completion and are not easily extinguished.

Hygiene Practices

- Keep your hands away from your face, eyes, mouth, and body while using chemicals.
- Food and drink, open or closed, should never be brought into the laboratory or chemical storage area.
- Never use laboratory glassware for eating or drinking purposes.
- Do not apply cosmetics while in the laboratory or storage area.
- Wash hands after removing gloves and before leaving the laboratory.
- Remove any protective equipment (i.e., gloves, lab coat or apron, chemical splash goggles) before leaving the laboratory.

Emergency Procedure

- Know the location of all the exits in the laboratory and building.
- Know the location of the emergency phone.
- Know the location of and know how to operate the following:
    - Fire extinguishers
    - Alarm systems with pull stations
    - Fire blankets
    - Eye washes
    - First-aid kits
    - Deluge safety showers
- In case of an emergency or accident, follow the established emergency plan as explained by the TA and evacuate the building via the nearest exit.

Chemical Handling

- Check the label to verify it is the correct substance before using it.

- Wear appropriate chemical resistant gloves before handling chemicals. Gloves are not universally protective against all chemicals.

- If you transfer chemicals from their original containers, label chemical containers as to the contents, concentration, hazard, date, and your initials.

- Always use a spatula or scoopula to remove a solid reagent from a container.

- Do not directly touch any chemical with your hands.

- Never use a metal spatula when working with peroxides. Metals will decompose explosively with peroxides.

- Hold containers away from the body when transferring a chemical or solution from one container to another.

- Use a hot water bath to heat flammable liquids. Never heat directly with a flame.

- Add concentrated acid to water slowly. Never add water to a concentrated acid.

- Weigh out or remove only the amount of chemical you will need. Do not return the excess to its original container, but properly dispose of it in the appropriate waste container.

- Never touch, taste, or smell any reagents.

- Never place the container directly under your nose and inhale the vapors.

- Never mix or use chemicals not called for in the laboratory exercise.

- Use the laboratory chemical hood, if available, when there is a possibility of release of toxic chemical vapors, dust, or gases. When using a hood, the sash opening should be kept at a minimum to protect the user and to ensure efficient operation of the hood. Keep your head and body outside of the hood face. Chemicals and equipment should be placed at least six inches within the hood to ensure proper air flow.

- Clean up all spills properly and promptly as instructed by the teacher.

- Dispose of chemicals as instructed by the TA.

- When transporting chemicals (especially 250 mL or more), place the immediate container in a secondary container or bucket (rubber, metal, or plastic) designed to be carried and large enough to hold the entire contents of the chemical.

- Never handle bottles that are wet or too heavy for you.

- Use equipment (glassware, Bunsen burner, etc.) in the correct way, as indicated by the TA.

WHAT IS A MATERIAL SAFETY DATA SHEET?

Material Safety Data Sheets (MSDS) contain information regarding the proper procedures for handling, storing, and disposing of chemical substances.

- An MSDS accompanies all chemicals or kits that contain chemicals.

- If an MSDS does not accompany a chemical, many web sites and science supply companies can supply one or they can be obtained from *www.msdsonline.com.*

- Typically the information is listed in a standardized format (ANSI Z400.1-1998, Hazardous Industrial Chemicals-Material Safety Data Sheet-Preparation).

- Refer to sections *General Guidelines to Follow in the Event of a Chemical Accident or Spill* and *Understanding an MSDS* in this front matter for additional information on the format and content of MSDSs (ANSI format).

COMMON SAFETY SYMBOLS

The United Nation's Globally Harmonized System (GHS) of classification and labeling of chemicals.

Health Hazard	Flame	Exclamation Mark

• Carcinogen • Mutagenicity • Reproductive toxicity • Respiratory sensitizer • Target organ toxicity • Aspiration toxicity	• Flammables • Pyrophorics • Self-heating • Emits flammable gas • Self-reactives • Organic peroxides	• Irritant (skin and eye) • Skin sensitizer • Acute toxicity (harmful) • Narcotic effects • Respiratory tract irritant • Hazardous to ozone layer (non-mandatory)
Gas Cylinder	**Corrosive**	**Exploding Bomb**

• Gases under pressure	• Skin corrosion/burns • Eye damage • Corrosive to metals	• Explosives • Self-reactives • Organic peroxides
Flame Over Circle	**Environment** (non-mandatory)	**Skull and Crossbones**

• Oxidizers	• Aquatic toxicity	• Acute toxicity (fatal or toxic)

NATIONAL FIRE PROTECTION ASSOCIATION HAZARD LABELS

The National Fire Protection Association (NFPA) has developed a visual guide (right) for a number of chemicals pertinent to the MSDS. The ANSI/NFPA 704 Hazard Identification system, the NFPA diamond, is a quick visual review of the health hazard, flammability, reactivity, and special hazards a chemical may present.

The diamond is broken into four sections (blue, red, yellow, and white). The symbols and numbers in the four sections indicate the degree of hazard associated with a particular chemical or material.

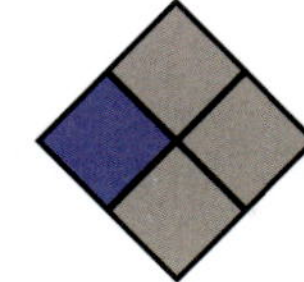

	Health Hazard (blue)	
4	Danger	May be fatal on short exposure. Specialized protective equipment required.
3	Warning	Corrosive or toxic. Avoid skin contact or inhalation.
2	Warning	May be harmful if inhaled or absorbed
1	Caution	May be irritating
0		No unusual hazard

	Flammability (red)	
4	Danger	Flammable gas or extremely flammable liquid
3	Warning	Combustible liquid flash point below 100°F
2	Caution	Combustible liquid flash point of 100° to 200°F
1		Combustible if heated
0		Not combustible

	Reactivity (yellow)	
4	Danger	Explosive material at room temperature
3	Danger	May be explosive if shocked, heated under confinement or mixed with water
2	Warning	Unstable or may react violently if mixed with water
1	Caution	May react if heated or mixed with water but not violently
0	Stable	Not reactive when mixed with water

	Special Notice Key (white)	
W	Water Reactive	
OX	Oxidizing Agent	

HOW DOES A CHEMICAL ENTER THE BODY?

A chemical can enter the body through different routes.

- These different routes of exposure and the types of exposure (acute or chronic) can affect the toxicity of the chemical.
- The most probable (primary) route(s) of exposure to a chemical will be identified in the MSDS.
- Three principal routes of exposure include dermal exposure (skin), inhalation, and ingestion (oral).

Dermal Exposure

Although the skin is an effective barrier for many chemicals, it is a common route of exposure. The toxicity of a chemical depends on the degree of absorption that occurs once it penetrates the skin. Once the skin is penetrated, the chemical enters the blood stream and is carried to all parts of the body. Chemicals are absorbed much more readily through injured, chapped, or cracked skin, or needle sticks than through intact skin. Generally, organic chemicals are much more likely to penetrate the skin than inorganic chemicals.

Dermal exposure to various substances can also cause irritation and damage to the skin and/or eyes. Depending on the substance and length of exposure, effects of dermal exposures can range from mild temporary discomfort to permanent damage.

Inhalation

Inhalation is another route of chemical exposure. Chemicals in the form of gases, vapors, mists, fumes, and dusts entering through the nose or mouth can be absorbed through the mucous membranes of the nose, trachea, bronchi, and lungs. Unlike the skin, lung tissue is not a very protective barrier against the access of chemicals into the body. Chemicals, especially organic chemicals, enter into the blood stream quickly. Chemicals can also damage the lung surface.

Ingestion

Ingestion involves chemicals entering the body through the mouth. Chemical dusts, particles, and mists may be inhaled through the mouth and swallowed.

They may also enter through contaminated objects, such as hands or food that come in contact with the mouth. Absorption of the chemicals into the bloodstream can occur anywhere along the length of the gastrointestinal (GI) tract.

WHAT ARE EXPOSURE LIMITS?

Exposure limits are intended to protect workers from excessive exposure to hazardous substances:

- Established by health and safety authorities and chemical manufacturers
    - Department of Labor's Occupational Safety and Health Administration (OSHA)
    - American Conference of Governmental Industrial Hygienists (ACGIH)
    - National Institute for Occupational Safety and Health (NIOSH)
    - Environmental Protection Agency (EPA)
    - American Industrial Hygiene Association (AIHA)
- Define the amount/concentration to which a worker can be exposed without causing an adverse health effect.
- Typically pertain to the concentration of a chemical in the air, but may also define limits for physical agents such as noise, radiation, and heat.
- Usually can be found on the MSDS; make sure your MSDSs are up-to-date.

Exposure Limits

Legally Enforceable Limits

Permissible Exposure Limits (PELs)

- Set by OSHA, 29 CFR 1910.1000, and 1910.1001 through 1910.1450.
- Specifies the maximum amount or concentration of a chemical to which a worker may be exposed.
- Generally defined in three different ways:
    - **Ceiling Limit (C):** The concentration that must not be exceeded at any part of the workday
    - **Short Term Exposure Limit (STEL):** The maximum concentration to which workers may be exposed for a short period of time (15 minutes)
    - **Time Weighted Average (TWA):** The average concentration to which workers may be exposed for a normal, 8-hour workday

Other U.S. Exposure Limits

Threshold Limit Values (TLVs)

- Prepared by ACGIH volunteer scientists
- Denotes the level of exposure that nearly all workers can experience without an unreasonable risk of disease or injury
- An advisory limit; not enforceable by law
- Generally can be defined as ceiling limits, short term exposure limits, and/or time-weighted averages
- Usually equivalent to PELs

Recommended Exposure Limits (RELs)

- Recommended by NIOSH
- Indicates the concentration of a substance to which a worker can be exposed for up to a 10-hour workday during a 40-hour work week without adverse effects, however, sometimes based on technical feasibility
- Based on animal and human studies
- Generally expressed as a ceiling limit, short-term exposure limit, or a time-weighted average
- Often more conservative than PELs and TLVs

Workplace Environmental Exposure Limits (WEELs)

- Developed by AIHA volunteers
- Advisory limits; not enforceable by law
- Typically developed for chemicals that are not widely used or for which little toxicity information is available

Company-Developed Limits

- Developed by company scientists
- Advisory limits; not enforceable by law
- Usually based on only short-term studies of animals
- Generally intended for internal company use and sometimes for the customers

GENERAL GUIDELINES TO FOLLOW IN THE EVENT OF A CHEMICAL ACCIDENT OR SPILL

- Assess the overall situation.
- Determine the appropriate action to resolve the situation.
- Follow the pre-existing, approved local emergency plan.
- Act swiftly and decisively.

Below are some recommended actions for specific emergencies. Some of the actions have been proposed by the Council of State Science Supervisors in Science and Safety: Making the Connection.

Chemical in the Eye

- Flush the eye immediately with water while holding the eye open with fingers.
- If wearing contact lens, remove and continue to rinse the eye with water.
- Continue to flush the eye and seek immediate medical attention.

Acid/Base Spill

For a spill not directly on human skin, do the following:

- Neutralize acids with powdered sodium hydrogen carbonate (sodium bicarbonate/baking soda), or bases with vinegar (5% acetic acid solution).
- Avoid inhaling vapors.
- Spread diatomaceous earth to absorb the neutralized chemical.
- Sweep up and dispose of as hazardous waste.

For spills directly on human skin, do the following:

- Flush area with copious amounts of cold water from the faucet or drench shower for at least 5 minutes.
- If spill is on clothing, first remove clothing from the skin and soak the area with water as soon as possible.
- Arrange treatment by medical personnel.

UNDERSTANDING AN MSDS

ANSI Standardized MSDS Format

Section 1 gives details on *what the chemical or substance is, CAS number, synonyms, the name of the company issuing the data sheet,* and often an *emergency contact number.*

Section 2 identifies the *OSHA hazardous ingredients,* and may include *other key ingredients* and exposure limits.

Section 3 lists the major *health effects* associated with the chemical. Sometimes both the acute and chronic hazards are given.

Section 4 provides *first aid measures* that should be initiated in case of exposure.

Section 5 presents the *fire-fighting measures* to be taken.

Section 6 details the *procedures to be taken in case of an accidental release.* The instructions given may not be sufficiently comprehensive in all cases, and local rules and procedures should be utilized to supplement the information given in the MSDS sheet.

Section 7 addresses the *storage and handling* information for the chemical. This is an important section as it contains information on the flammability, explosive risk, propensity to form peroxides, and chemical incompatibility for the substance. It also addresses any special storage requirements for the chemical (i.e., special cabinets or refrigerators).

Section 8 outlines the *regulatory limits for exposure,* usually the maximum permissible exposure limits (PEL) (refer to section *How Does a Chemical Enter the Body?* of the front matter). The PEL, issued by the Occupational Safety and Health Administration, tells the concentration of air contamination a person can be exposed to for 8 hours a day, 40 hours per week over a working lifetime (30 years) without suffering adverse health effects. It also provides information on personal protective equipment.

Section 9 gives the *physical and chemical properties* of the chemical. Information such as the evaporation rate, specific gravity, and flash points are given.

Section 10 gives the *stability and reactivity* of the chemical with information about chemical incompatibilities and conditions to avoid.

Section 11 provides both the *acute and chronic toxicity* of the chemical and any health effects that may be attributed to the chemical.

Section 12 identifies both the *ecotoxicity* and the environmental fate of the chemical.

Section 13 offers suggestions for the *disposal of the chemical.* Local, State, and Federal regulations should be followed.

Section 14 gives the *transportation information* required by the Department of Transportation. This often identifies the dangers associated with the chemical, such as flammability, toxicity, radioactivity, and reactivity.

Section 15 outlines the *regulatory information* for the chemical. The hazard codes for the chemical are given along with principle hazards associated with the chemical. A variety of country and/or state specific details may be given.

Section 16 provides *additional information* such as the label warnings, preparation and revision dates, name of the person or firm that prepared the MSDS, disclaimers, and references used to prepare the MSDS.

SAMPLE MSDS

Material Safety Data Sheet

Toluene   *MSDS No. XXXX*

1. Product and Company Identification

Product Name: Toluene

Synonyms: Methylbenzene, Methylbenzol, Phenylmethane, Toluol

CAS No.: 108–88–3

Chemical Formula: C_6H_5–CH_3

Catalog Number: Tol 12

Supplier: Company X

        XXXXXXXXXX

        Anywhere, XX XXXXX

        Emergency Information: 800-XXX-XXXX

2. Composition/Information on Ingredients

Ingredient	CAS No	Percent	Hazardous
Toluene	108–88–3	100%	Yes

3. Hazards Identification

Emergency Overview

Danger! Harmful or fatal if swallowed. Vapor harmful. Poison! May be absorbed through intact skin. Flammable liquid and vapor. May cause liver and kidney damage, may affect blood system or central nervous system. Causes irritation to skin, eyes, and respiratory tract.

Potential Acute Health Effects

Eye Contact: Causes severe eye irritation with redness and pain.

Skin Contact: Causes irritation. May be absorbed through skin.

Inhalation: Inhalation may cause irritation of the upper respiratory tract. Symptoms of overexposure may include fatigue, confusion, headache, dizziness, and drowsiness. Very high concentrations may cause unconsciousness and death.

Ingestion: Swallowing may cause abdominal spasms and other symptoms that parallel over-exposure from inhalation. Aspiration of material into the lungs may cause chemical pneumonitis, which may be fatal.

Chronic Exposure: Chronic exposure may result in anemia, decreased blood cell count, and bone marrow hypoplasia. Liver and kidney damage may occur. Repeated or prolonged contact may cause dermatitis.

4. First Aid Measures

Eye Contact: Immediately flush eyes with plenty of water for at least 15 minutes, lifting the upper and lower eyelids occasionally. Get medical attention immediately.

Skin Contact: In case of contact, immediately flush skin with plenty of soap and water for at least 15 minutes while removing contaminated clothing and shoes. Wash clothing before reuse. Call a physician immediately.

Inhalation: Evacuate victim to fresh air immediately. If not breathing, give artificial respiration. If breathing is difficult, give oxygen. Seek medical aid immediately.

Ingestion: Aspiration hazard. If swallowed, ***do not induce vomiting.*** Give 2–4 cups of milk or water. Never give anything by mouth to an unconscious person. Get medical attention immediately.

5. Fire Fighting Measures

Fire: Flash point: 4°C (40°F)

Autoignition temperature: 480°C (896°F)

Flammable limits in air % by volume: lower: 1.3%; upper: 7.1%

Flammable liquid and vapor!

Extremely flammable when exposed to flame or sparks. Vapors are heavier than air and can flow along surfaces to distant ignition source and flash back.

Explosion: Vapor-air concentrations above flammable limits are explosive. Contact with strong oxidizers may cause fire or explosion. Sensitive to static discharge.

Fire Extinguishing Media: Dry chemical, carbon dioxide, or foam. Material is lighter than water and a fire may be spread by use of water. Water may be used to cool fire surface and protect personnel. Water may also be used to flush spills away from exposures and to dilute spills to non-flammable mixtures. Avoid flushing hydrocarbon into sewers.

Special Information: In the event of a fire, wear full protective clothing and NIOSH-approved self-contained breathing apparatus operated in the pressure demand or other positive pressure mode.

6. Accidental Release Measures

Avoid Contact: Ventilate area of leak or spill. Remove all ignition sources. Wear appropriate personal protective equipment as specified in Section 8 Isolate hazard area. Contain and recover liquid when possible. Collect liquid in an appropriate container or absorb with an inert material such as earth, sand, or vermiculite. Do not use combustible materials, such as saw dust. Do not flush to sewer.

7. Handling and Storage

Handling: Wash thoroughly after handling. Use with adequate ventilation. Avoid contact with skin, eyes, or clothes. Electrically ground and bond containers when transferring material to avoid static accumulation.

Storage: Store in a cool, dry, well-ventilated location, away from any area where the fire hazard. Separate from incompatibles. Storage and use areas should be No Smoking areas. Use non-sparking type tools and equipment, including explosion proof ventilation. Containers of this material may be hazardous when empty since they retain product residues (vapors, liquid). Observe all warnings and precautions listed for the product. Protect container against physical damage. Keep container tightly closed.

8. Exposure Controls/Personal Protection

Ventilation System: A system of local and/or general exhaust is recommended to keep exposures below the Airborne Exposure Limits.

Exposure Limits: Toluene:

- OSHA Permissible Exposure Limit (PEL): 200 ppm TWA; 300 ppm (acceptable ceiling conc.); 500 ppm (acceptable maximum conc.)
- NIOSH Recommended Exposure Limit (REL): 100 ppm TWA (375 mg/m^3); STEL 150 ppm (560 mg/m^3)
- ACGIH Threshold Limit Value (TLV): 50 ppm TWA skin—potential for cutaneous absorption

Personal Respirators (NIOSH/EN 149 Approved): If the exposure limit is exceeded a half-face organic vapor respirator may be worn for up to ten times the exposure limit. A full-face organic vapor respirator or self-contained breathing apparatus may be worn up to 50 times the exposure limit. For emergencies or instances where the exposure levels are not known, use a full-face piece positive-pressure, air-supplied respirator.

Skin Protection: Wear impervious protective clothing, including boots, gloves, lab coat, apron or coveralls, as appropriate, to prevent skin contact.

Eye Protection: Use chemical splash goggles and/or a full face shield. Maintain eyewash fountain facilities in work area.

9. Physical and Chemical Properties

Physical State and Appearance: Clear, colorless liquid

Odor: Aromatic benzene-like

Solubility: Very slight

Specific Gravity (Water = 1): 0.9

Viscosity: 20cP @ 20°C

Boiling Point: 110°C (232°F)

Melting Point: −95°C (−139°F)

Vapor Density (Air = 1): 3.1

Vapor Pressure (mmHg): 53.3 @ 20°C (68°F)

Evaporation Rate (Butyl acetate = 1): 2.4

Molecular Formula: $C_6H_5CH_3$

Molecular Weight: 92.06

10. Stability and Reactivity

Stability: Stable under ordinary conditions of use and storage. Containers may burst when heated.

Hazardous Decomposition Products: Carbon dioxide and carbon monoxide may form when heated to decomposition.

Hazardous Polymerization: Has not been reported.

Incompatibilities: Heat, flame, strong oxidizers, nitric and sulfuric acids; will attack some forms of plastics, rubber, coatings.

Conditions to Avoid: Heat, flames, ignition sources and incompatibles.

11. Toxicological Information

Toxicological Data

Oral rat LD_{50}: 636 mg/kg

Skin rabbit LD_{50}: 14100 uL/kg

Inhalation rat LC_{50}: 49 gm/m^3/4H

Inhalation mouse LC_{50}: 400 ppm/24H

Irritation data: skin rabbit, 500 mg, Moderate

Eye rabbit, 2 mg/24H, Severe

Investigated as a tumorigen, mutagen, reproductive effector.

Reproductive Toxicity

Has shown some evidence of reproductive effects in laboratory animals.

12. Ecological Information

Environmental Fate: When released into the soil, this material may evaporate and is micro biologically biodegradable. When released into the soil, this material is expected to leach into groundwater. When released into water, this material may evaporate and biodegrade to a moderate extent. When released into the air, this material may be moderately degraded by reaction with photochemically produced hydroxyl radicals.

Environmental Toxicity: No data available; however, this material is expected to be toxic to aquatic life.

13. Disposal Considerations

Waste material should be handled as hazardous waste and sent to a RCRA approved incinerator or disposed in a RCRA approved waste facility. Processing, use, or contamination of this product may change the waste management options. State and local disposal regulations may differ from Federal disposal regulations. Dispose of container and unused contents in accordance with Federal, State, and local requirements.

14. Transport Information

Domestic (Land, U.S. D.O.T.)

Proper Shipping Name: Toluene

Hazard Class: 3

UN/NA: UN1294

Packing Group: II

Canada TDG

Proper Shipping Name: Toluene

Hazard Class: 3 (9.2)

UN/NA: UN1294

Packing Group: II

Additional Information: Flashpoint 4 C

15. Regulatory Information

CALIFORNIA PROPOSITION 65: WARNING

This product contains a chemical known to the State of California to cause birth defects or other reproductive harm.

Reportable Quantity: 1000 Pounds (454 Kilograms) (138.50 Gals)

NFPA Rating: Health – 2; Fire – 3; Reactivity – 0 (0 = Insignificant; 1 = Slight; 2 = Moderate; 3 = High; 4 = Extreme)

Carcinogenicity Lists: No

NTP: No

IARC Monograph: No

OSHA Regulated: No

Section 313 Supplier Notification: This product contains the following toxic chemical(s) subject to the reporting requirements of SARA TITLE III Section 313 of the Emergency Planning and Community Right-To-Know Act of 1986 and of 40 CFR 372:

CAS No.	Chemical Name	% by Weight
108–88–3	Toluene	100

16. Other Information

Label Hazard Warning

POISON! DANGER! HARMFUL OR FATAL IF SWALLOWED. HARMFUL IF INHALED OR ABSORBED THROUGH SKIN. VAPOR HARMFUL. FLAMMABLE LIQUID AND VAPOR. MAY AFFECT LIVER, KIDNEYS, BLOOD SYSTEM, OR CENTRAL NERVOUS SYSTEM. CAUSES IRRITATION TO SKIN, EYES, AND RESPIRATORY TRACT.

Label Precautions

- Keep away from heat, sparks, and flame. Keep container closed.
- Use only with adequate ventilation. Wash thoroughly after handling.
- Avoid breathing vapor.
- Avoid contact with eyes, skin, and clothing.

Label First Aid

Aspiration hazard. If swallowed, ***do not induce vomiting.*** Give large quantities of water. Never give anything by mouth to an unconscious person. If vomiting occurs, keep head below hips to prevent aspiration into lungs. If inhaled, remove to fresh air. If not breathing, give artificial respiration. If breathing is difficult, give oxygen. In case of contact, immediately flush eyes or skin with plenty of water for at least 15 minutes. Remove contaminated clothing and shoes. Wash clothing before reuse. In all cases call a physician immediately.

References: Upon request

Molecular Models
Organic Compounds

Learning Objectives

- Understand and explore the bonding characteristics of organic compounds.
- Identify organic structures that are structural isomers, geometric isomers, or optical isomers of one another.
- Recognize chiral carbons.
- Draw and build chemically correct organic structures.

Additional Reading and References

- Timberlake: 12.1, 12.2, 12.3, 12.5, 12.6
- Timberlake: 15.2

PRE-LAB QUESTIONS

Name: ________________________________   Partners: ________________________________

TA: ________________________________   Section: ______________   Date: ______________

Answer the following questions using information obtained from lecture, the textbook, or lab manual. The responses will be checked at the start of your lab section for credit. Failure to complete may result in exclusion from participating in the lab.

1. Define the following terms in a few words:

a. isomers:

b. structural isomers:

c. geometric isomers:

d. optical isomers (enantiomers):

2. What is the requirement for a carbon to be considered *chiral?*

3. How many bonds can/should be drawn to the following atoms in an organic structure?

   a. carbon: **c.** hydrogen: **e.** bromine, chlorine:

   b. oxygen: **d.** nitrogen:

4. Properly draw a molecule that would be considered to have

   (Show all hydrogens and approximately correct bond angles.)

   a. a *trans* double bond.

   b. a *cis* double bond.

BACKGROUND

Isomers

In order for two molecules to be isomers of each other, they must have the same molecular formula; that is, the same number of C, H, and other atoms. If they do not have exactly the same number and kinds of atoms, they are simply different molecules. They may be part of a series of very similar molecules which can be discussed as a group, such as a series of hydrocarbons, but without identical molecular formulas they are not isomers. There are two main types of isomers to consider: structural isomers and enantiomers.

Structural Isomers

Structural isomers are those molecules which have the same molecular formula but a different sequence of atoms and bonds. As an example, note the following two structures for C_4H_8:

$$CH_2\!\!=\!\!CH-CH_2-CH_3 \qquad\qquad CH_3-CH\!\!=\!\!CH-CH_3$$

$$\textbf{A} \qquad\qquad\qquad\qquad\qquad \textbf{B}$$

Molecules A and B are structural isomers because the sequence of atoms and bonds is different between the two molecules. There are other possible structural isomers of C_4H_8. One of them is shown below as structure **C.**

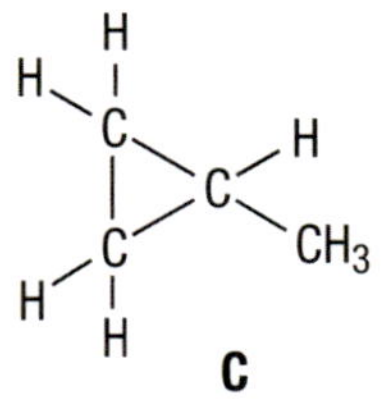

$$\textbf{C}$$

Obviously, the sequence of atoms and bonds in **C** is different from that in either **A** or **B**.

As proficiency is gained in organic nomenclature, a simple way to determine whether two molecules are structural isomers is to name both of them. If the same name is generated, then they are identical, not structural isomers.

When a different atom, such as oxygen, is introduced into the molecule, other structural isomers are possible. A pair of isomers for the formula C_2H_6O is shown below as molecules **D** and **E.**

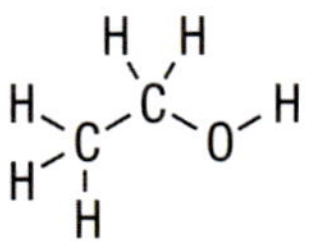

$$\textbf{D} \qquad\qquad\qquad\qquad \textbf{E}$$

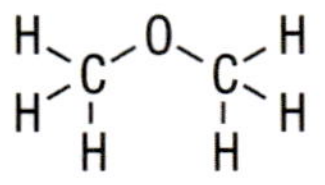

Again the different sequence of bonds is evident. **D** could be classified as an alcohol, while **E** would be an ether.

As shown in Table 1.1, the number of structural isomers of saturated hydrocarbons increases slowly at first and then more rapidly as the number of carbon atoms is increased. When two H atoms are removed from the molecules, making them unsaturated, the number of possible bonding arrangements increases more rapidly. When a non-carbon atom, such as oxygen, is added to the formula of even a saturated hydrocarbon, the number of isomers increases very rapidly.

TABLE 1.1 Structural Isomers for Saturated, Unsaturated Structures, and Saturated with One Oxygen That Contain Eight or Less Carbons						
	Saturated Molecules		**Unsaturated Molecules***		**Saturated with 1 Oxygen**	
Number of Carbons	**Formula**	**Number of Isomers**	**Formula**	**Number of Isomers**	**Formula**	**Number of Isomers**
1	CH_4	1	—	—	CH_4O	1
2	C_2H_6	1	C_2H_4	1	C_2H_6O	2
3	C_3H_8	1	C_3H_6	2	C_3H_8O	3
4	C_4H_{10}	2	C_4H_8	5	$C_4H_{10}O$	7
5	C_5H_{12}	3	C_5H_{10}	10	$C_5H_{12}O$	14
6	C_6H_{14}	5	C_6H_{12}	25	$C_6H_{14}O$	32
7	C_7H_{16}	9	C_7H_{14}	56	$C_7H_{16}O$	72
8	C_8H_{18}	18	C_8H_{16}	140	$C_8H_{18}O$	>>150

These include cyclic molecules, as well as those with a double bond.

Before continuing on to stereoisomers, it is useful to pause and consider the molecular formulas for saturated and unsaturated hydrocarbons. In looking at the columns in Table 1.1 associated with saturated compounds containing no oxygens (or any other heteroatoms), a general relationship between the number of carbons and number of hydrogens found in a molecule may be observed:

Number of Hydrogens = Twice the Number of Carbons + Two
or
Number of Hydrogens = 2n + 2 (where n is the number of carbons in the compound)

So, if a compound contains no double or triple bonds (or rings), the molecular formula will be of the form C_nH_{2n+2} and the compound may be classified as an alkane.

A similar pattern results when the molecular formulas of the series of unsaturated molecules is examined. Notice that in each of these, a pair of hydrogens has been removed to generate what is referred to as a single point of unsaturation (a double bond or a ring). Again, a general relationship between the number of carbons and the number of hydrogens found in each of these compounds is easily generated:

Number of Hydrogens = Twice the Number of Carbons
or
Number of Hydrogens = 2n (where n is the number of carbons in the compound)

So, if there is a double bond or ring in the molecule, the molecular formula will be of the form C_nH_{2n} and the compound is either a cycloalkane or a monoalkene.

A similar pattern could be found for those containing two points of unsaturation (like compounds containing a triple bond or alkynes). The general molecular formula for an alkyne is C_nH_{2n-2}. Based on Table 1.1, what is the general formula for a saturated compound that contains one oxygen? (ANS: $C_nH_{2n+2}O$) Similar general molecular formulas can be found for other common families of organic compounds.

Stereoisomers

Another large group of molecules that differ from each other, but are not structural isomers of one another, are stereoisomers. Stereoisomers are molecules which have the same molecular formula and the same sequence of atoms and bonds, but different arrangements in space. There are several sub-categories of stereoisomers, but we will focus on only two of them—geometric isomers and optical isomers (enantiomers).

(I) Geometric Isomers (*cis-trans* isomers)

Geometric isomers are those molecules which have the same molecular formula and the same sequence of atoms and bonds, but different arrangements in space, due to double bonds or rings. Illustrating the effect of double bonds are molecules **F** and **G** which are two different versions of molecule **B**.

The sequence of atoms and bonds is the same in both **F** and **G**: a CH_3 attached to a $C{=}C$ attached to a CH_3. These two molecules are not structural isomers of each other, but their atoms differ in their arrangement in space, so **F** and **G** are geometric isomers. The arrangement in **F** which has both CH_3 groups extending on the same side of the double bond is called the ***cis*** geometric isomer; whereas the arrangement in **G** which has the CH_3 groups extending on opposite sides of the double bond is called the ***trans*** geometric isomer.

(II) Enantiomers (optical isomers)

Enantiomers, like geometric isomers, are types of stereoisomers. Recall that stereoisomers are defined as molecules which have the same molecular formula and the same sequence of atoms and bonds, but different arrangements in space. Enantiomers are pairs of stereoisomers which are mirror images of each other, but are not superimposable on top of each other.

Although there are a few exceptions for some special types of compounds, in our study of organic compounds there will have to be at least one carbon in the molecule with four different groups attached to it in order for enantiomers to exist. Such a carbon is called a **chiral carbon** (pronounced with a hard "k" and a long "i" sound).

Enantiomers occur due to the tetrahedral geometry about a carbon atom which has four single bonds attached to it. (Thus, carbons that have a double bond attached cannot be enantiomers as they do not meet this geometric requirement.) Since the tetrahedral shape is not encountered frequently in everyday life, one important part of this experiment is to spend time constructing and studying models which have lots of tetrahedral shapes connected to one another.

If you can mentally visualize three dimensions, it is easily observed that molecules **H** and **I** are mirror images of each other, but they are not superimposable on top of each other. If this is not obvious, use the models in lab, and ask fellow students or the laboratory instructor to help you see it.

   H   H
   | Cl   Cl |
  C   C
F   Br   Br   F
   H   **I**

The term **optical isomers** comes from the fact that when a beam of polarized light is passed through a collection of molecules of one enantiomer, like **H,** the plane of polarization of the light is rotated. The rotation of plane polarized light is an optical effect; hence, the term optical isomers. Although the term optical isomers is used loosely to mean enantiomers, it is not exact because there are other types of stereoisomers that also rotate plane polarized light.

In order to show enantiomers (with their tetrahedral shapes) on a flat piece of paper, it is necessary to use some sort of representation of a three-dimensional structure. One way to do this is to use wedge-shaped lines which imply that the atom attached to the wide end is extending toward you (in front of the page), and dotted lines which imply that the atom attached to one end is extending away from you (behind the page). Ordinary lines represent bonds lying in the plane of the paper. For example, in molecule **H** the H, C, and F atoms lie in the plane of the paper, while the bromine is in front of the paper (toward you) and the chlorine is behind the paper (away from you). This represents a tetrahedral arrangement of atoms. In the case of molecule **I,** the H, C, and F atoms lie in the plane of the paper, while the bromine is in front of the paper (toward you) and the chlorine is behind the paper (away from you).

PROCEDURE

Use the guidance that follows to complete the subsequent worksheet pages. Build as many models as necessary to understand the concepts and achieve the stated learning objectives. You may work together in groups and discuss the answers with your teaching assistant to make sure you understand each part prior to moving to the next exercise.

1. Begin with building enantiomers to focus attention on the tetrahedral shape of carbon which has four single bonds attached to it.

   a. Construct the two enantiomers of the molecule, bromochloroiodomethane (CHClBrI). Convince yourself that they are mirror images but are not superimposable.

   b. Draw a representation of each molecule, similar to the ones shown as **H** and **I**. Ask your laboratory instructor to check to see that they are correct. Make sure to draw the molecules on your paper such that they are close to the actual 109.5° observed.

TABLE 1.2 Molecular Model Kits	
Molecular model kits usually presume certain color conventions for atoms, as follows.	
Model Color	**Atom**
Black	Carbon
Green	Chlorine
White	Hydrogen
Orange	Bromine
Red	Oxygen
Purple	Iodine
Blue	Nitrogen
Yellow	Sulfur

When molecules composed of tetrahedral carbon atoms are drawn in two dimensions without using wedge-shaped and dotted lines, no stereochemistry is implied by the drawing; all that is being shown by the drawing is the connections between atoms.

2. a. Construct a model of ethene (also known as ethylene), C_2H_4. (The longer, flexible bonds should be used for constructing the bonds associated with the double bond.)

   b. Draw a representation of this molecule, and include the bond angles in your picture. Please note that, as in this molecule, whenever a carbon atom has one double bond and two single bonds attached to it, the C atom and the three attached atoms all lie in the same plane.

3. a. Construct a model of ethyne (also known as acetylene), C_2H_2.

   b. Draw a representation of this molecule, and include the bond angles in your picture. Please note that whenever a carbon atom has one triple bond and one single bond attached to it, the C atom and the two attached atoms all lie in a straight line.

4. a. Construct the two geometric isomers of 1,2–dichloro–1–propene ($ClCH=CClCH_3$).

   b. Draw representations of the two molecules and label them as *cis* or *trans*.

5. To see whether you have things straight in your mind, examine each pair of molecules shown on the worksheet, and decide whether the two molecules of the pair are **different molecules, structural isomers, identical molecules, geometric isomers,** or **enantiomers.**

(They can only be *one* of the above. If you have any trouble with them—especially the enantiomers—build models to help you decide.)

6. The four structures shown below are merely different ways of writing the molecule, 2,3–dimethylbutane, C_6H_{14}. These pictures do not represent isomers; they all represent the same molecule. If you have any difficulty seeing that these are the same molecule, build models and/or discuss them with other students or the laboratory instructor.

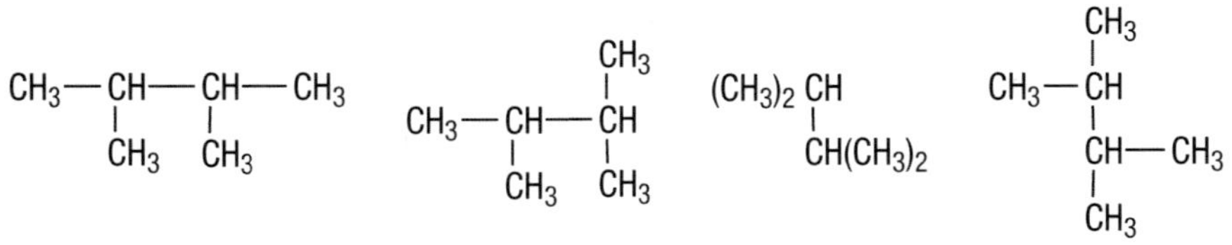

Now for the grand finale to this experiment, draw the structures for all the possible isomers of C_4H_{10}, then of C_4H_8, and finally of $C_4H_{10}O$. These are found in row four of Table 1.1, so you have the advantage of knowing when you have identified all the isomers (or when you haven't!).

Prior to the end of class, have your Data Sheet signed by your TA. Complete any remaining work as the Post-Lab assignment to be turned in at the start of the next lab period.

DATA SHEET

Name: _________________________________   Partners: _________________________________

TA: _________________________________   Section: _____________   Date: _____________

1. Draw a representation of each **CHClBrI** molecule, similar to the ones shown on page 28.

2. Draw a representation of ethene, C_2H_4, and include the bond angles in your picture.

3. Draw a representation of ethyne, C_2H_2, and include the bond angles in your picture.

4. Draw representations of the two geometric isomers of **1,2–dichloro–1–propene** and label them as *cis* or *trans.*

5. Examine each of the following pairs of molecules and decide whether the two molecule pairs are **different molecules, structural isomers, identical molecules, geometric isomers,** or **enantiomers.**

a.

$CH_3CHCH_2CH_3$
   |
   CH_3

CH_3
   |
CH_3CHCH_2
      |
      CH_3

b.

H_3C CH_2CH_3
      C
H_3C CH_2CH_3

H_3C CH_3
      C
H_3C CH_2CH_3

c.

H_3C CH_3
      C
H_3C CH_2CH_3

H_3C CH_2CH_3
      C
H_3C CH_3

d.

H   H
 C=C
Cl CH_2CHCH_3
        |
        CH_3

H   H
 C=C
Cl $CHCH_3$
       |
       CH_3

e.

H_3C OH
      C
H CH_2CH_3

H   OH
   C
H_3C CH_2CH_3

f.

$CH_3CH_2CH_2$ H
           C=C
   H CH_2CHCH_3
          |
          CH_3

H_3C
   |
CH_3CHCH_2 H
           C=C
   H $CH_2CH_2CH_3$

g.

      CH_2
H_2C CH_2
H_2C—CH_2

$CH_3CH_2CH_2CH=CH_2$

h.

CH_3 CH_2CH_3
      C=C
Cl H

CH_3CH_2 Cl
        C=C
   H CH_3

i.

         CH_3
          |
CH_3—CH—CH
       | |
      CH_3 CH_3

CH_3—CH—CH_2—CH_2
       | |
      CH_3 CH_3

Name: ___________________________________   Partners: ___________________________________

TA: _____________________________________   Section: _______________   Date: _______________

6. Draw the structures for all possible isomers of the molecules given below.

C_4H_{10}

C_4H_8

$C_4H_{10}O$

<table>
<tr><td>

TA Signature

</td></tr>
</table>

Pre-Lab Assignment ________	Ask your TA to review your work and sign your report. The TA will
Safety/Participation ________	sign above once satisfied that the student has performed the entire
Lab Write-Up ________	procedure. The report will not be accepted or graded unless signed.

Dyes and Dyeing

Learning Objectives

- Recognize common organic functional groups.

- Understand the characteristics of dye molecules and the importance of conjugation.

- Make connections between the fastness of a dye and the type or strength of molecular interactions.

- Record and collect experimental data and observations for the purpose of writing a lab report.

- How to write a scientific lab report that includes the following sections: introduction, experimental, results and discussion, and conclusions.

Additional Reading and References

- Reisch, M. S., Better Times Ahead for U.S. Dye Producers. *Chemical & Engineering News* **1988,** 66(30), 7–14.

Safety Precautions and Hazards

- Chemical splash goggles and lab coat must be worn at all times.

- Nitrile gloves must be worn when handling chemicals.

- Use tongs or clothes pins, rather than fingers, to handle dyed pieces of cloth.

PRE-LAB QUESTIONS

Name: _______________________________   Partners: _______________________________

TA: _________________________________   Section: ______________   Date: _______________

Answer the following questions using information obtained from lecture, the textbook, or lab manual. The responses will be checked at the start of your lab section for credit. Failure to complete may result in exclusion from participating in the lab.

1. Identify the circled functional groups within the molecules below and give the common name for each dye:

a.

b.

Common name:

c.

d.

e.

Common name:

f.

g.

h.

Common name:

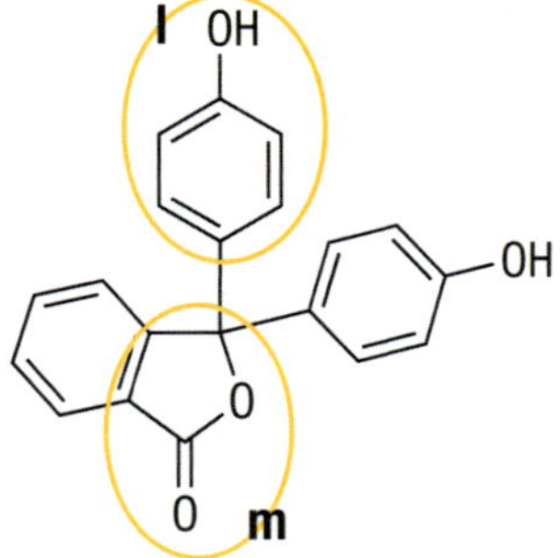

i.

j.

k.

Common name:

l.

m.

Common name:

n.

o.

Common name:

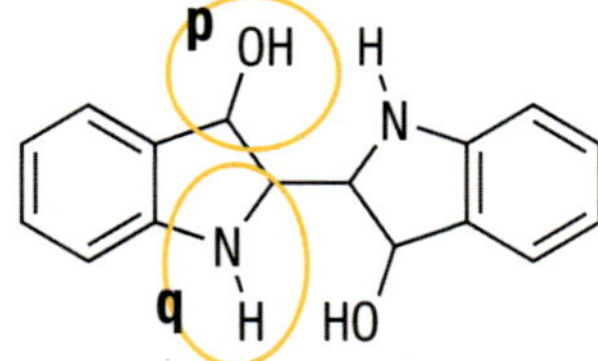

p.

q.

Common name:

r.

s.

Common name:

2. The alternation of a single and double bonds in organic compounds is often referred to as *conjugation.* Based on your readings, how does extent of conjugation effect the color of a molecule? (Refer to the visible electromagnetic spectrum in your answer.)

3. Explain what is meant by color fastness and how a **_mordant_** additive might aid in achieving this.

BACKGROUND

When a molecule absorbs light, the energy of the absorbed light may be transferred to the molecule by promoting an electron from one molecular orbital to a higher energy orbital. The less firmly an electron is held to the molecular frame work in its initial state, the easier it will be to raise it to a higher energy orbital. It is very often the case that non-bonding electrons, or unshared pairs, such as those found on N or O and those electrons found in double or triple bonds are relatively easy to promote to a higher energy level. When there are appropriate energy levels fairly close in energy, the light absorbed to achieve this promotion is in the ultraviolet or visible range of the electromagnetic spectrum (Figure 2.1). Notice how the color of light changes based on the wavelength.

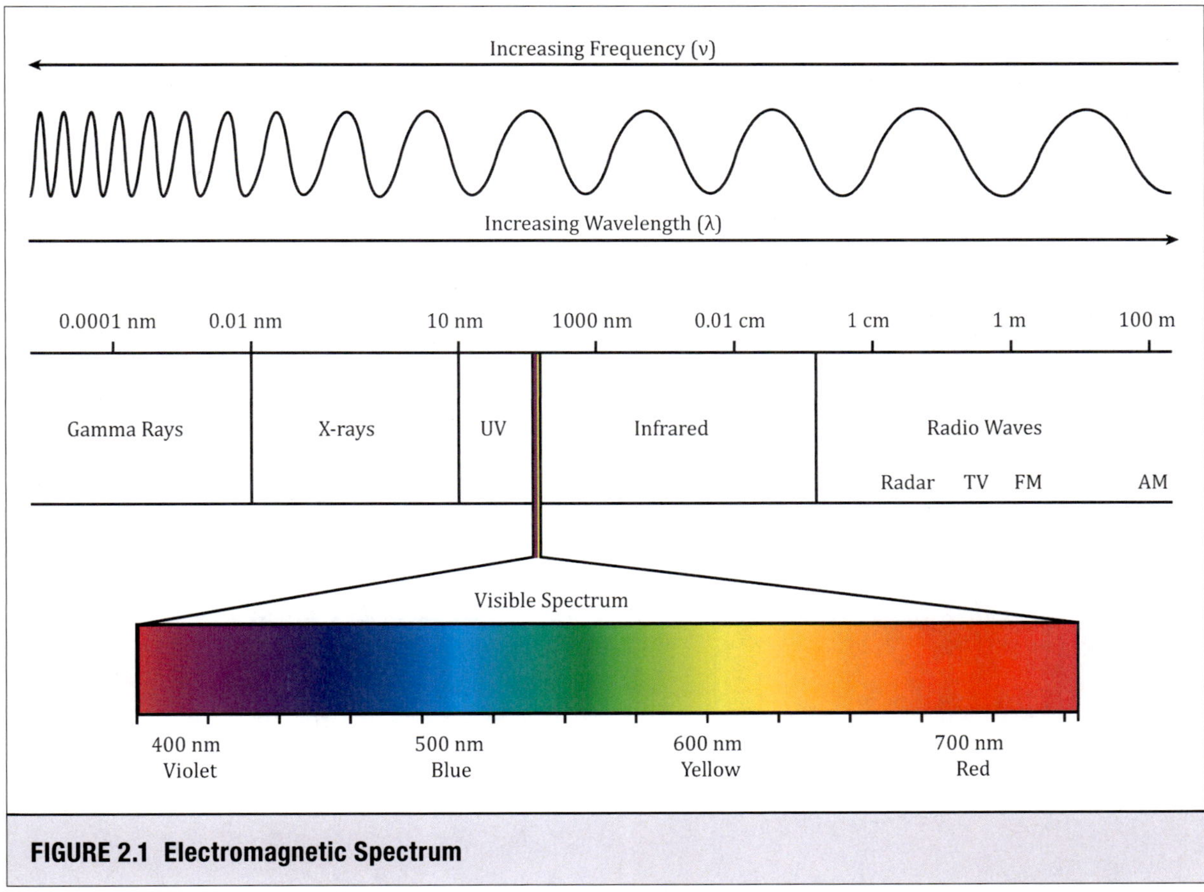

FIGURE 2.1 Electromagnetic Spectrum

Dye molecules are colored because they partially absorb light in the visible region of the spectrum. Light absorption is made possible by the presence of **chromophores** (from Greek, *chroma* meaning "color" and *phoros* meaning "bearer"). Chromophores are molecules or parts of molecules that contain unsaturated groups such carbon-carbon double bonds ($C\!=\!C$) but also functionality like azo ($—N\!=\!N—$), nitro ($O\!=\!N—O$), carbonyl ($C\!=\!O$), or alkynyl ($C\!\equiv\!C$) groups. By themselves, each of these groups would absorb in the ultraviolet region of the spectrum. For example, simple azo compounds are colorless, absorbing at 340 nm. But, when these groups are connected in such a way that they form a system that is referred to as **conjugated,** the energy required to excite an electron to a higher level is decreased and the molecule will absorb in the visible region of the spectrum. For instance, azobenzene (Figure 2.2) is orange and absorbs in the visible at 445 nm. Most dyes have extended conjugated systems (Figure 2.4).

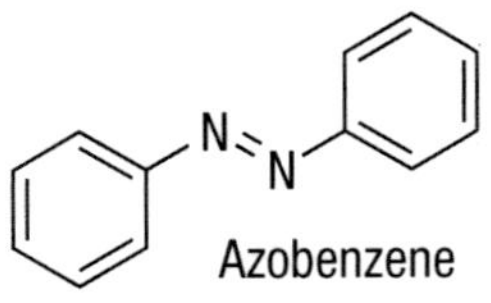

FIGURE 2.2 Structure of Azobenzene

If one were to compare the structural characteristics of the molecules in Figure 2.4, what is the defining feature that makes them all considered to be chromophores? Notice how all of these molecules have extensive regions of alternating single and double bonds. It is this pattern that is referred to as conjugation. The greater amount of conjugation found in a molecule, the further the absorption of the molecule is shifted to the red end of the visible spectrum. Molecules that have little conjugation are going to tend to absorb in the ultraviolent region.

Dyes also usually contain **auxochromes** (from Latin, *auxilium* meaning aid) which includes groups such as amino ($-NH_2$, $-NR_2$), phenolic ($-OH$), alkoxyl ($-OR$), sulfonyl ($-SO_3H$), and carboxyl ($-CO_2H$). When these are suitably attached to a conjugated system, they tend to increase the wavelength at which light is absorbed and intensify the absorption. Auxochromes also frequently serve as anchoring groups, helping to bind the dye through either chemical covalent bonds or via intramolecular hydrogen-bonding with functional groups on the fiber.

Dyes may be separated into broad classes (see referenced article for the complete list). Depending upon the type of functional groups present in the molecule, they will interact with very different materials. In this particular experiment, the focus is on dyes used primarily with fabrics. The chemical formulas for the two dyes to be used in this experiment follow in Figure 2.3. Note how their structures fulfill the requirements outlined above for conjugated molecules.

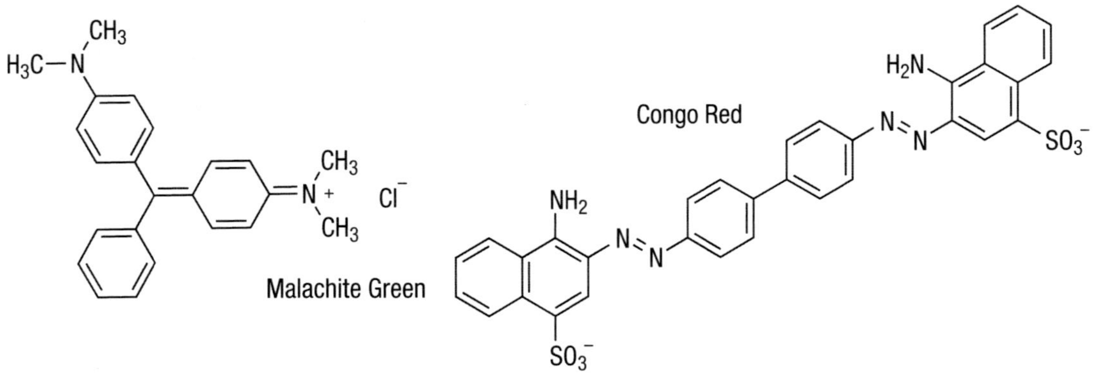

FIGURE 2.3 Structures of Malachite Green and Congo Red

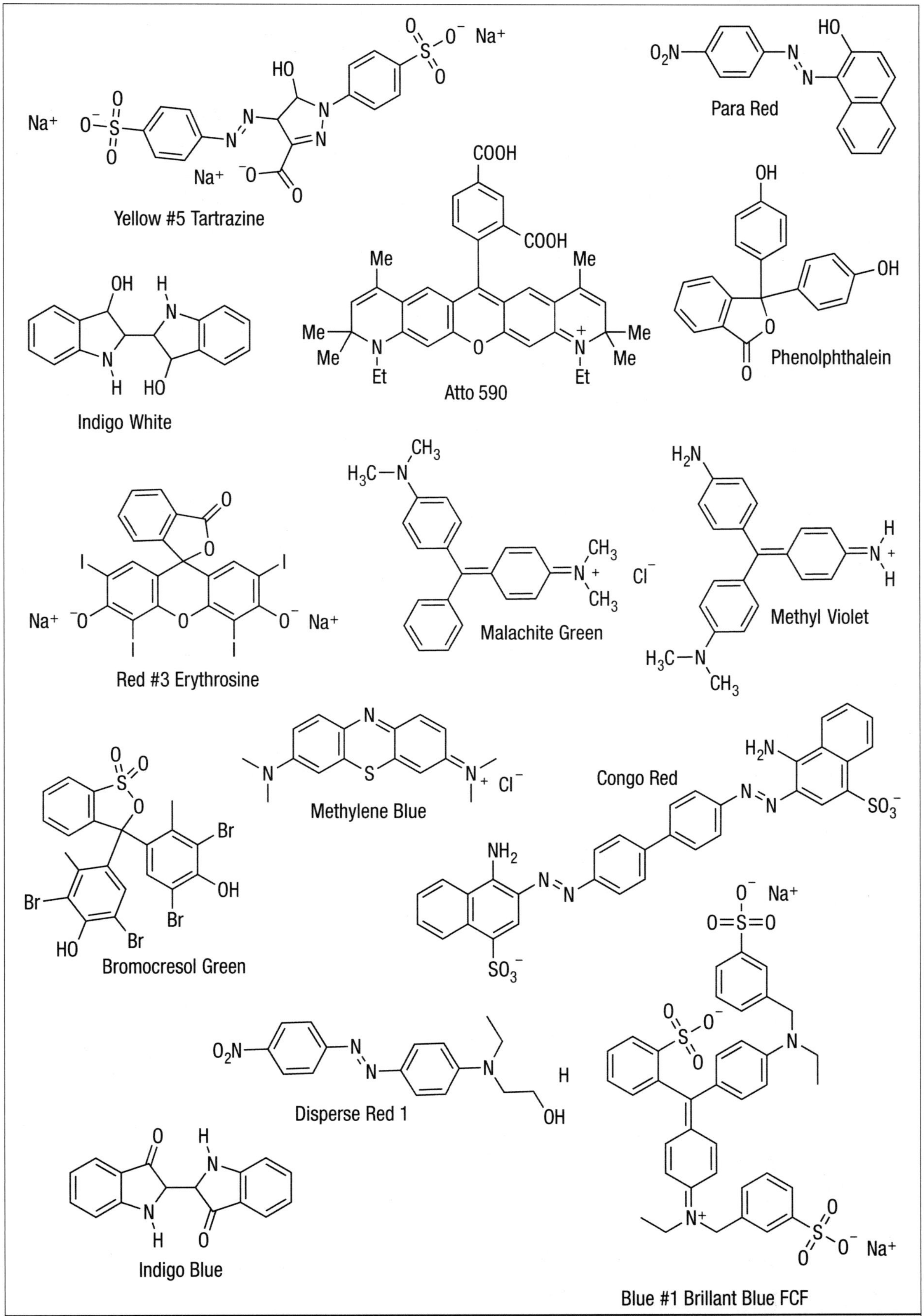

FIGURE 2.4 Various Molecules with Extended Conjugated Systems

Direct and fiber-reactive dyes adhere to cloth without the aid of supplementary chemicals. Wool and silk both contain many anionic (negatively charged) polar sites which readily bind to positively charged dyes such as malachite green or methyl violet. The first satisfactory direct dye for cotton was Congo Red. It has polar amine and sulfonate groups which can hydrogen-bond to the hydroxyl groups in cotton (or other dye molecules), thus making it less susceptible to removal by washing. Water-soluble, anionic dyes contain a reactive group that can form a covalent bond with a compatible group on the cloth surface. Such covalent bonds may be formed with the hydroxyl groups of cellulosic fibers such as cotton and rayon or with the amino groups in nylon, silk, and wool.

Some dye molecules are unable to dye cloth directly due to their inability to suitably interact with the chemical functional groups. They can often be made to do so, however, with the aid of a **mordant.** A mordant is a binding link between the fiber and the dye; the mordant is first fixed to the fabric, and the dye then binds to the mordant. Often, mordants are metallic hydroxides which form salts, or they may chelate with the fiber and with the dye. Tannins, albumin, and other polar substances may also be used as mordants. In this experiment, the use of tannic acid as a mordant with Malachite Green will illustrate their effect (Figure 2.5). Most mordants are mainly used to dye wool, but the class has secondary applications for dyeing silk, nylon, leather, and even anodized aluminum. The use of mordants was one of the first methods to produce wash-fast fabrics.

FIGURE 2.5 Structure of Tannic Acid

PROCEDURE

A. Congo Red

1. Using the concentrated stock solution provided in the hood, add approximately two full pipettes of Congo Red solution to ~75 mL water in a beaker.

2. To this solution, add a second solution containing 0.2 g sodium carbonate that has been dissolved in 10 mL water.

**Congo Red causes serious eye irritation.
It is also teratogenic and carcinogenic.**

3. Slowly heat combined Congo Red and sodium carbonate solution to boiling.

4. Place a piece of either wool, silk, or cotton in the hot solution for 2 minutes.

5. Remove the cloth piece and make some initial observations about the color.

6. Rinse using water for 2 minutes and record observations.

7. Repeat Steps 4–6 with the remaining fabric samples. If the color of the solution has noticeably faded since the start of the experiment, add an additional pipette full of stock Congo Red solution. (**Note:** If you are so interested, you could also run a set of experiments where the length of time is varied to see if that impacts the extent of fastness.)

B. Malachite Green

8. Using the concentrated stock solution provided in the hood, add approximately 1–2 full pipettes of malachite green solution to ~200 mL water in a beaker.

Malachite green is harmful if swallowed and causes serious eye damage. It is also a suspected teratogen and toxic to aquatic life.

9. Slowly heat the solution to boiling.

10. Place a piece of either wool, silk, or cotton in the hot solution for 2 minutes.

11. Remove the cloth piece and make some initial observations about the color.

12. Rinse using water for 2 minutes and note any changes in color.

13. Repeat Steps 10–12 with the remaining fabric samples. If the color of the solution has noticeably faded since the start of the experiment, add an additional pipette full of stock malachite green solution. (**Note:** If you are so interested, you could also run a set of experiments where the length of time is varied to see if that impacts the extent of fastness.)

14. Save the malachite green solution for next part.

C. Malachite Green with a Mordant

15. Dissolve 0.2 g tannic acid in 100 mL of water in a separate beaker.

16. Slowly heat the solution to boiling.

17. Immerse a strip of either wool, silk, or cotton in the tannic acid solution for about 1 minute.

18. Remove the cloth and press the tannic acid solution from it using a spatula. (This can perhaps be most easily done inside another clean beaker.)

19. Place the piece of cloth in the boiling malachite green solution for 2 minutes.

20. Remove the cloth piece and make some initial observations about the color.

21. Rinse using water and note any changes in color.

22. Repeat Steps 17–21 with the remaining fabric samples. If the color of the solution has noticeably faded since the start of the experiment, add an additional pipette full of stock malachite green solution. (**Note:** If you are so interested, you could also run a set of experiments where the length of time is varied to see if that impacts the extent of fastness.)

DATA SHEET

Name: _______________________________   Partners: _______________________________

TA: _______________________________   Section: _____________   Date: _____________

Dye	Material	Observations
Congo Red	Wool	
Congo Red	Silk	
Congo Red	Cotton	
Malachite Green	Wool	
Malachite Green	Silk	

Dye	Material	Observations
Malachite Green	Cotton	
Malachite Green + Tannic Acid	Wool	
Malachite Green + Tannic Acid	Silk	
Malachite Green + Tannic Acid	Cotton	

Conclusions:

TA Signature

Pre-Lab Assignment ________
Safety/Participation ________
Lab Write-Up ________

Ask your TA to review your work and sign your report. The TA will sign above once satisfied that the student has performed the entire procedure. The report will not be accepted or graded unless signed.

POST-LAB

The post-lab assignment for this experiment is a lab report. Broad directions are provided in Appendix B of this lab manual. Your TA should have some additional suggestions or guidelines that specifically pertain to this experiment. As they are the ones who will be grading the report, it is important to follow their specific requirements.

What follows here are some ideas to consider as you are writing your reports:

- What the similarities and differences in how the fabrics interact with the dye molecules? Think about the different functional groups present in the dye molecules and the fabrics.

- How did the use of a mordant change these interactions?

- What are the interactions or bonds formed that are causing the varying degrees of fastness?

- How does the color of the dye relate to conjugation?

- What additional experiments would you like to perform to collect more data? Different dyes? Different fabrics?

Alcohols, Aldehydes, and Ketones

Learning Objectives

- Classify alcohols as primary, secondary, or tertiary.

- Predict the product of oxidation reactions involving organic functional groups.

- Understand the interplay of polar and nonpolar groups of aldehydes, ketones, and alcohols in determining the solubility of a molecule in water or hexanes.

- Visually recognize whether two liquids are miscible or immiscible.

Additional Reading and References

- Timberlake: 13.3, 13.4

- Timberlake: 14.2

Safety Precautions and Hazards

- Chemical splash goggles and lab coats must be worn at all times.

- Nitrile gloves must be worn when handling chemicals.

- All waste must be disposed in a waste container. No chemicals from this experiment can go down the drain.

PRE-LAB QUESTIONS

Name: _________________________________   Partners: _______________________________

TA: ___________________________________   Section: ______________   Date: ________________

Answer the following questions using information obtained from lecture, the textbook, or lab manual. The responses will be checked at the start of your lab section for credit. Failure to complete may result in exclusion from participating in the lab.

1. Complete the following table. Classify each substance as a 1°, 2°, or 3° alcohol, a ketone, or an aldehyde.

Name	Drawing	Functional Group Classification
Methanol		
2–Propanol		
1–Butanol		
1–Pentanol		
Benzaldehyde		
Acetone		
Cinnamaldehyde		

2. What determines whether a particular compound is soluble in hexanes versus water? (Talk about polarity and intermolecular interactions in your answer.)

3. What are some of the indicating factors that a chemical reaction has occurred?

BACKGROUND

One of the most fundamental classes of reactions in all of chemistry is oxidation-reduction reactions. The broadest definition of an oxidation-reduction reaction is a chemical transformation where there is a transfer of electrons occurring. These reactions are responsible for many modern inventions we take for granted. For instance, batteries work due to the oxidation-reduction reactions occurring with two metal electrodes. If we assume for the sake of example that one of the electrodes is copper and the other is zinc, the electron-transfer is more easily discerned:

$$Cu^{2+}(aq) + Zn(s) \rightarrow Zn^{2+}(aq) + Cu(s)$$

In this case, zinc is being oxidized from the 0 oxidation state to the 2+ oxidation state, while at the same time copper is being reduced from 2+ to 0. This change in oxidation state is due to the fact that two electrons are being transferred from the zinc metal to the copper ions. Oxidation always occurs in tandem with a reduction. We can describe the reactant that is *causing* the oxidation of another species as the oxidizing agent. The species that *causes* the reduction of another species is the reducing agent.

These principles that were established in an introductory chemistry course still apply to organic reactions, but often the approach to recognizing these reactions is simplified so that oxidation states need not be strictly observed. For organic reactions, oxidation occurs when either of two things happen: (1) there is a loss of two hydrogens **or** (2) there is an increase in number of carbon-oxygen bonds. Very often, both of these are observed in the same reaction. There are numerous reactions in organic chemistry that would fit this broad description, however, the focus for this experiment will be on the transformations of alcohols, aldehydes, and ketones. If you examine the relationship between the functional groups (Figure 3.1), it is clear that aldehydes and ketones are the result of oxidation reactions involving certain types of alcohols.

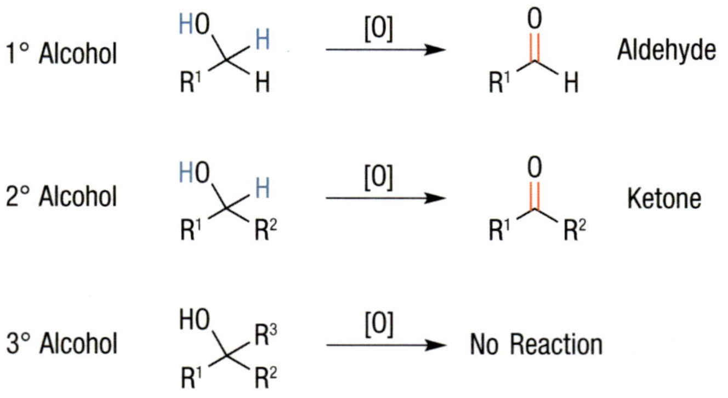

FIGURE 3.1 Oxidation of Alcohols [O] is a generic oxidizing agent; R^1, R^2, and R^3 are all alkyl or aryl groups.

Alcohols may be classified according to the number of hydrogens attached to the carbon bearing the hydroxyl group. If there are two hydrogens attached (and only one carbon group), the alcohol is classified as a primary alcohol (1°) and the product of oxidation is an aldehyde. If there is only one hydrogen attached (and two carbon groups), the alcohol is classified as a secondary alcohol (2°) and the product of oxidation is a ketone. In each of these reactions shown in Figure 3.1, note that there is a loss of two hydrogens and a gain in carbon-oxygen bonds, thus meeting the classification requirement for an oxidation reaction. Also, notice that if the alcohol is a tertiary alcohol (3°), with no hydrogens attached to the carbon bearing the hydroxyl group, there is no reaction when the alcohol is mixed with an oxidizing agent.

It is also worth considering the potential for further oxidizing aldehydes and ketone by again removing two hydrogens or adding carbon-oxygen bonds. From examining the reactions of alcohols, a fundamental requirement for an oxidation to occur is that the carbon being oxidized must bear at least one hydrogen. This leads to the prediction that aldehydes may be further oxidized and ketones cannot. Figure 3.2 presents the likely results.

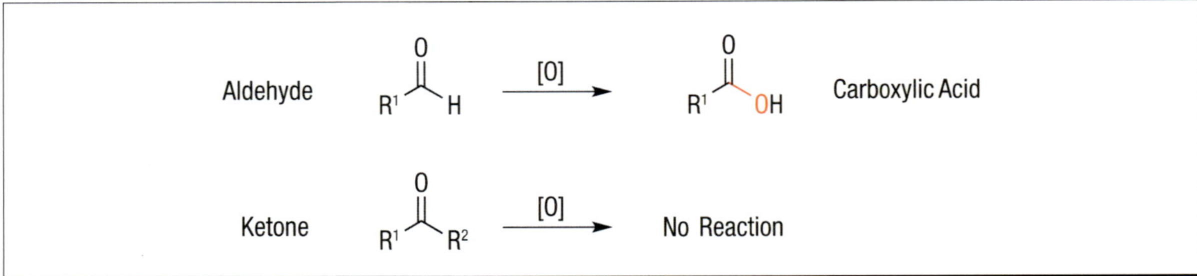

FIGURE 3.2 Oxidation of Aldehydes and Ketones [O] is a generic oxidizing agent; R^1 and R^2 are alkyl or aryl groups.

Aldehydes will be further oxidized to make carboxylic acids, while ketones will not undergo further reaction with the oxidant. It is often the case with most oxidizing agents that a primary alcohol will be fully oxidized all the way to a carboxylic acid with no possibility of stopping the reaction at the aldehyde.

This experiment explores not only the chemical properties of alcohols, aldehydes, and ketones, but also their physical properties. The chemical properties of these functional groups will be tested by exposing them to an oxidizing agent (potassium permanganate, $KMnO_4$). The solubility of alcohols with varying numbers of carbon atoms will be tested using both polar and nonpolar solvents. Similarly, the solubility of an aldehyde and a ketone will also be tested. How might you expect the solubility of these to differ from the alcohols? As you are performing these solubility tests, think about the polar and nonpolar regions of each alcohol, aldehyde, or ketone and the kinds of intermolecular forces that may be governing their overall physical properties. After exploring the physical and chemical properties, conclusions will be drawn about the various functional groups.

PROCEDURE

A. Oxidation of Alcohols, Aldehydes, and Ketones

In this test, an oxidizing agent ($KMnO_4$), will be added to various organic compounds to test their ability to be oxidized. Since the reactions can take some time to occur, this test should be started first, but data should not be recorded until the other sections have been performed.

1. Label test tubes (or a test tube rack) with the six compounds to be tested:

   #1. water (control)   **#3.** 2–propanol   **#5.** benzaldehyde

   #2. ethanol   **#4.** 2–methyl–2–propanol   **#6.** acetone

2. Add approximately half a pipette of each compound to the appropriate test tube.

3. Add one drop of the $KMnO_4$ solution to each test tube.

4. Allow the test tubes to sit while performing the other sections of the experiment.

5. Once the other sections of the experiment have been performed record the color of the solution in each test tube.

The organic compounds used in this experiment are volatile and very flammable. They are harmful if swallowed and cause skin irritation. Do not breathe vapors.

B. Solubility in Hexanes

In this test, compounds will be mixed with hexanes to determine their solubility in a non-polar solvent.

6. Start with six clean test tubes. Label the test tubes (or rack) with the six compounds to be tested:

   a. water   **c.** ethanol   **e.** benzaldehyde

   b. methanol   **d.** 2–propanol   **f.** acetone

7. Add approximately half a pipette of the respective compounds to each test tube.

8. Add approximately half a pipette of hexane to each test tube.

9. Observe if the compound and the hexanes mix completely, or if the compounds separate. If a line can been seen dividing the hexanes from the compound, the compound is not soluble in the hexanes. Consult with your TA should there be any question as to the miscibility of the two liquids.

10. Shake the test tube to see if the compound separates from the hexanes even after being agitated.

11. Record your results in the data table.

12. Rinse your test tubes and dispose of all waste in the appropriate container.

13. Save your test tubes for reuse in Part C.

C. Solubility in Water

In this test, various compounds will be mixed with water to determine their solubility in a polar solvent.

14. Start with the same six clean test tubes used in Part B.

15. Label the test tubes (or a test tube rack) with the six compounds to be tested:

i. methanol   **iii.** 2–propanol   **v.** benzaldehyde

ii. ethanol   **iv.** 1–butanol   **vi.** acetone

16. Fill each test tube approximately ¼ of the way full with water.

17. Add approximately half a pipette full of each compound to the test tubes i–vi.

18. Observe if the compound and water mix completely, or if the compounds separate. If a line can be seen dividing the water from the compound, the compound is not soluble in water. Consult with your TA should there be any question as to the miscibility of the two liquids.

19. Shake the test tube to see if the compound separates from the water even after being agitated.

20. Record your results in the data table.

21. Rinse your test tubes and dispose of all waste in the appropriate container.

<table>
<tr><td>SAFETY!

WASTE

DISPOSAL</td><td>All waste must be disposed of in a waste container.
No chemicals from this lab can go down the sink.</td></tr>
</table>

D. Aromatic Compounds

Determine the odor of benzaldehyde and cinnamaldehyde by wafting the compound toward your nose as demonstrated by the TA.

DATA SHEET

Name: _________________________________   Partners: _________________________________

TA: _________________________________   Section: _____________   Date: _____________

A. Oxidation of Alcohols, Aldehydes, and Ketones

Compound	Color of Test Tube	Oxidized or Not Oxidized?	Draw Oxidation Product (put X if no product and explain why)
#1 – Water (control)			
Comments:			
#2 – Ethanol			
Comments:			
#3 – 2–Propanol			
Comments:			
#4 – 2-Methyl-2-Propanol (tert–butyl alcohol)			
Comments:			
#5 – Benzaldehyde			
Comments:			
#6 – Acetone			
Comments:			

B. Solubility in Hexanes

Compound	a. Water	b. Methanol	c. Ethanol	d. 2–Propanol	e. Benzaldehyde	f. Acetone
Soluble in hexanes?						
Comments:						

C. Solubility in Water

Compound	i. Methanol	ii. Ethanol	iii. 2–Propanol	iv. 1–Butanol	v. Benzaldehyde	vi. Acetone
Soluble in water?						
Comments:						

D. Aromatic Compounds

1. Describe the odor of benzaldehyde:

2. Describe the odor of cinnamaldehyde:

TA Signature

Pre-Lab Assignment _________ Safety/Participation _________ Lab Write-Up _________	Ask your TA to review your work and sign your report. The TA will sign above once satisfied that the student has performed the entire procedure. The report will not be accepted or graded unless signed.

POST-LAB

Conclusions

1. Based on the data collected, which functional groups can be oxidized and which cannot?

2. Based on the data collected, what factor seems to determine if a compound is soluble in water? Explain.

3. Based on the data collected, which factor seems to determine if a compound is soluble in hexanes? Explain.

Questions

1. Use your drawing of the structure of acetone from the pre-lab to answer the following questions:

   a. Why did acetone show no reaction with the oxidizing agent?

   b. Why is acetone soluble in both the hexanes and in water?

2. Based on results from this experiment, give a general rule that describes solubility of organic compounds in water. Be sure to include a discussion of intermolecular interactions.

Identification of an Unknown Carbohydrate

Learning Objectives

- Recognize the structural differences between monosaccharides, disaccharides, and polysaccharides.
- Understand what it means to be reducing sugar.
- Using experimental data, make a reasonable identification of an unknown carbohydrate.
- Make a connection between the fundamental organic functional groups and the reactions that can be performed on carbohydrates.

Additional Reading and References

- Timberlake: 15.1–15.7

Safety Precautions and Hazards

- Chemical splash goggles and lab coat must be worn at all times.
- Nitrile gloves must be worn when handling chemicals.
- Pay attention to waste disposal notes in the procedure.

PRE-LAB QUESTIONS

Name: _______________________________________   Partners: _______________________________________

TA: ____   Section: _______________   Date: _______________

Answer the following questions using information obtained from lecture, the textbook, or lab manual. The responses will be checked at the start of your lab section for credit. Failure to complete may result in exclusion from participating in the lab.

1. Identify each of these carbohydrates as either **glucose, fructose,** or **galactose.** Classify each using the notation described for monosaccharides.

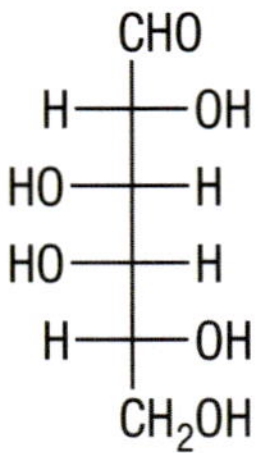 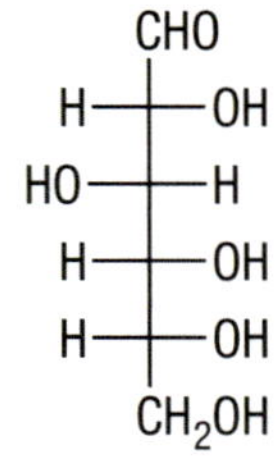 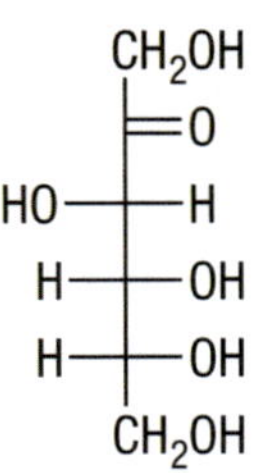

2. In a few words, describe what will be observed for a positive test using each of these reagents:

Benedict's reagent:

Barfoed's solution:

Seliwanoff's reagent:

Iodine test:

3. What are the differences in glycosidic linkages between **amylose, amylopectin,** and **glycogen?**

BACKGROUND

Carbohydrates are polyhydroxyl compounds that contain either an aldehyde or ketone functional group. The name carbohydrate refers to the approximate formula $C(H_2O)_n$ and highlights the major role the alcohol groups have in the physical and chemical properties. In addition to their functional groups, carbohydrates may be classified by the number of carbon atoms they contain and by the bonds formed with other sugars. As a class of carbohydrates, the simplest sugars are called monosaccharides. Other common sugars, like those involving two monosaccharides linked together, are called disaccharides. When many sugar molecules are linked together to form a polymer, this class of carbohydrate is referred to as polysaccharides.

Monosaccharides are classified by the number of carbon atoms present (5 = pentose, 6 = hexose) as well as the functional group present (aldehyde = aldose, ketone = ketose). The most common monosaccharides contain six carbons and either an aldehyde or ketone and therefore are classified as either aldohexoses or ketohexoses. Notice in Figure 4.1 that common monosaccharides glucose and galactose are both aldohexoses while fructose is a ketohexose.

Disaccharides are classified according to their monosaccharide composition as well as the type of bond connecting the two sugars together. Disaccharides are formed from a condensation reaction between the hemiacetal of a monosaccharide and a hydroxyl group of a second monosaccharide. The result is the formation of an acetal group. The new bond that connects the two molecules is an ether bond called a **glycosidic bond.** When this bond is formed using the hydroxyl groups from the hemiacetals of both sugars, it cannot be oxidized and the carbohydrate is called a **nonreducing sugar.** If only one of the hemiacetals is involved in the glycosidic bond and the other remains unreacted, the disaccharide is available to be oxidized and is thus called a **reducing sugar.** In Figure 4.2, for lactose there remains a hemiacetal that is not part of the glycosidic bond, however for sucrose both of the carbons from the original hemiacetals are participating in the glycosidic bond. Therefore, lactose is a reducing sugar and sucrose is a nonreducing sugar.

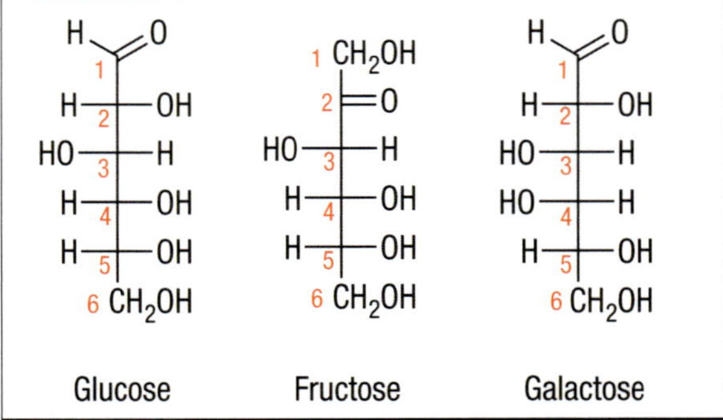

FIGURE 4.1 Fischer Projections of Glucose, Fructose, and Galactose

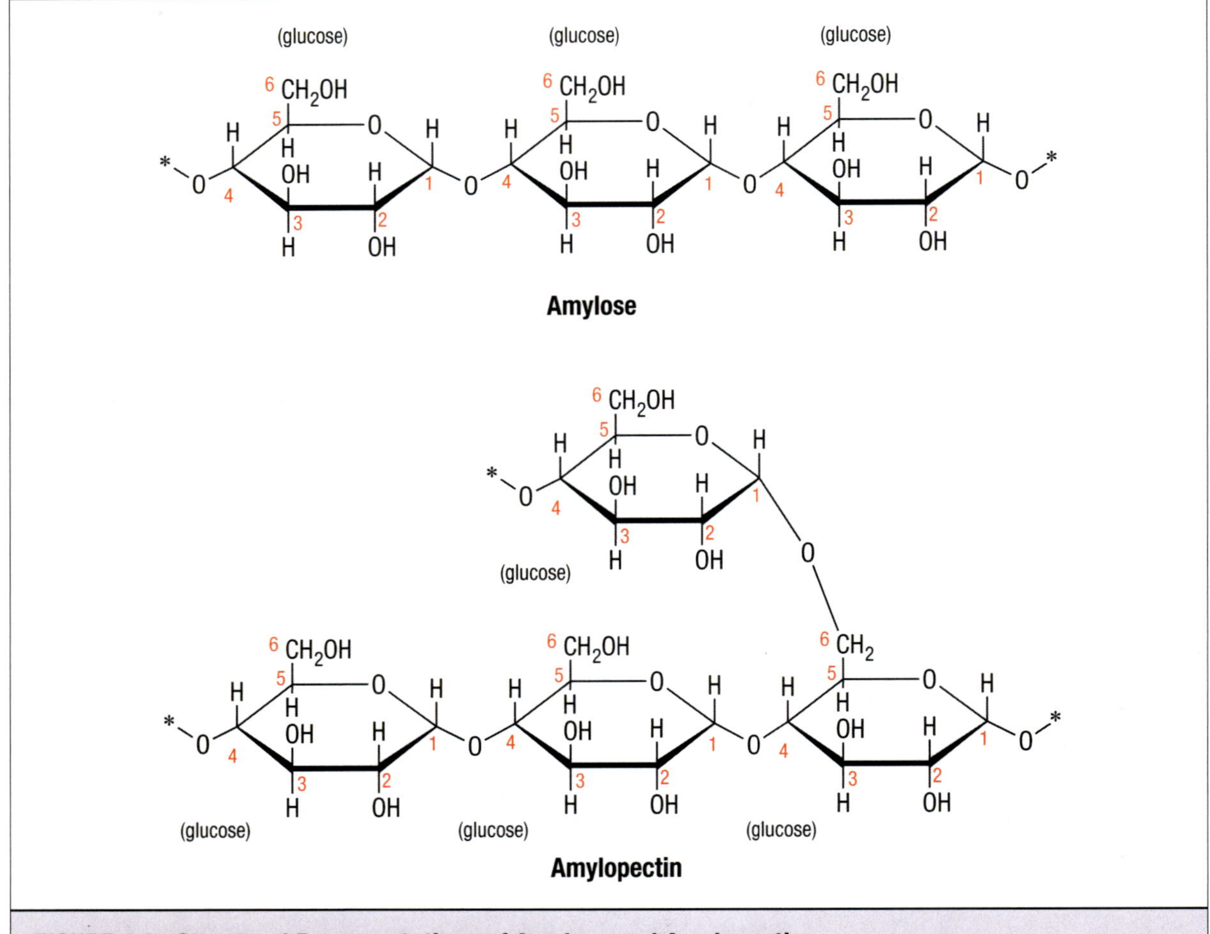

FIGURE 4.2   Haworth Structures of Lactose and Sucrose

Polysaccharides are composed of many sugar molecules linked together via glycosidic bonds. Starches and cellulose are both polysaccharides composed entirely of glucose molecules. The only difference between starch and cellulose is the type of linkage between the glucose molecules. Starches can be further classified as amylose (linear plant starch), amylopectin (branched plant starch), and glycogen (branched animal starch).

FIGURE 4.3   Structural Representations of Amylose and Amylopectin

In this experiment, a series of qualitative tests will be performed on carbohydrates from these various classes. The results of these tests will then be used to draw conclusions about an unknown carbohydrate. The tests to be performed in this experiment should help to identify reducing sugars (Benedict's reagent), monosaccharides (Barfoed's solution), keto-hexoses (Seliwanoff's reagent), and the presence of starch (iodine). Descriptions of each test follow below:

Qualitative Tests for Carbohydrates

Benedict's reagent tests for the presence of **reducing sugars.** Benedict's reagent is an alkaline (basic) solution containing Cu^{2+} ions that oxidize aldehydes. Because ketone groups cannot be further oxidized, it might be predicted that ketoses would be unreactive toward Benedict's reagent. However, in basic solution, ketoses actually isomerize into aldoses which are subsequently able to be oxidized. The end result is that all monosaccharides and disaccharides that are reducing sugars test positive with Benedict's reagent. The formation of Cu_2O as a yellow, orange, or brick red precipitate is the visual signifier of a positive test. This reagent is often used in clinical settings to test for glucose in the urine of diabetics.

Barfoed's solution is used to test for the presence of **monosaccharides.** The solution has the same blue color as the Benedict's reagent due to the presence of copper ions. Barfoed's solution, however, is acidic and under these conditions monosaccharides undergo rapid oxidation, while disaccharides are significantly slower to oxidize. If heated briefly, only monosaccharides display a positive test as evidence by the formation of a red Cu_2O precipitate.

Seliwanoff's reagent is used to test for the presence of ketoses. **Ketohexoses** dehydrate in the presence of 6 M hydrochloric acid to form hydroxymethyl-furfural (Figure 4.4). This product may then condense with resorcinol to give an orange or reddish product. Monosaccharides that are ketohexoses themselves or disaccharides that containing a ketohexose will both generate a red product. Other sugars will yield products ranging in color from yellow to light pink.

HO OH

Resorcinol

H O O OH

Hydroxymethylfurfural

FIGURE 4.4 Structures of Resorcinol and Hydroxylmethylfurfural

Iodine is used to test for the presence of **starch.** The iodine is typically mixed with potassium iodide to increase solubility in water. The initial solution should be a brownish-orange color that changes to dark blue or black in the presence of starch. The solution will not react with simple sugars such as mono- or disaccharides.

PROCEDURE

The following procedures are written such that each test is performed on all the sugars at the same time. Pay careful attention to directions for each specific test. Some require merely "warm" water, while others need boiling water. If possible, devote a hot plate solely for warming and another for boiling. There are going to be many different test tubes used in this experiment. It is important that you clearly label each test tube that you are using.

A. Enzymatic Behavior of Saliva on Starch

Before beginning the test, prepare a water bath using a beaker that is small enough that the test tubes can rest inside without spilling their contents. Fill the beaker half full with water and place on a hot plate set to a warm, but not boiling. Watch this beaker carefully to ensure the water does not boil off.

In this test, saliva will be mixed with starch and allowed to react for 30 minutes. After the reaction is complete, the starch/saliva mixture will be tested with iodine and Benedict's reagent to identify the sugars that remain in the test tube.

1. Begin the test by collecting roughly 2 mL of saliva in a clean test tube.

2. Add approximately the same volume of distilled water as saliva to the test tube and agitate to mix well.

3. Pour half of the saliva mixture into a second clean test tube.

4. Label these test tubes S1 and S2.

5. Add 10 drops of starch solution to each of the test tubes and allow to sit while completing the rest of the experiment.

After at least 30 minutes have passed:

6. Add three drops of Benedict's reagent to the test tube labeled S1.

Benedict's reagent causes serious eye irritation and is toxic to aquatic life.

7. Agitate to mix well and place the test tube labeled S1 in the **warm water bath** for several minutes. Record the results.

8. Add one drop of iodine solution to the test tube labeled S2. Agitate to mix well and record the results.

B. Benedict's Test for Reducing Sugars

9. Label test tubes for each of the five known carbohydrates (#1 glucose, #2 fructose, #3 sucrose, #4 lactose, #5 starch), one for a water comparison (#6), and one for your unknown carbohydrate (#7).

10. Gently shake each of the stock carbohydrate solutions, then add one pipette of each carbohydrate solution into the respective test tubes. Add one pipette full of water to test tube #6.

11. Add three drops of Benedict's reagent to each test tube and agitate to mix thoroughly.

Benedict's reagent causes serious eye irritation and is toxic to aquatic life.

12. Place the test tubes into the **warm water bath.**

13. Observe the test tubes for at least 2 minutes noting any color changes.

14. Remove the test tubes after 3 minutes and empty the test tubes into the appropriate waste container. Benedict's reagent cannot go down the drain.

C. Barfoed's Test for Monosaccharides

15. Refill the respective test tubes (including the water comparison) with two pipettes worth of each carbohydrate solution. Make sure the labels have not been washed or wiped off.

16. Add a full pipette of Barfoed's reagent to each test tube and agitate to mix.

CAUTION — **Barfoed's reagent contains copper(II) acetate, which is moderately toxic.**

17. Place all test tubes into the **boiling water bath** and heat for 2 minutes.

18. Observe the test tubes for at least 2 minutes noting any color changes.

19. Empty the test tubes down the sink with plenty of running water.

D. Seliwanoff's Test for Ketohexoses

20. Make sure the boiling water bath is still halfway full of water. Check that the labels have not been washed or wiped off of the test tubes.

21. To the empty test tubes, add a full pipette of Seliwanoff's reagent to each test tube.

Seliwanoff's reagent contains hydrochloric acid, which causes severe skin burns and eye damage.

22. Add one drop of each carbohydrate solution (or distilled water) to the appropriate test tube. Agitate to mix well.

23. Place all test tubes into the **boiling water bath** and heat for 3 minutes.

24. Note any color changes and record observations in the data table.

25. Empty the test tubes down the sink with plenty of running water.

26. Turn off the hot plate for the boiling water.

E. Iodine Test for Starches

27. Add two pipettes worth of each carbohydrate solution or distilled water to the appropriate test tubes.

28. Add one drop of iodine solution and agitate to mix well.

29. Record the results of the tests in the data table.

30. Empty the test tubes down the sink with plenty of running water.

31. If you have not done so already, go back and finish Part A.

DATA SHEET

Name: _________________________________   Partners: _________________________________

TA: ___________________________________   Section: ______________   Date: ________________

Data Table

Carbohydrate	Benedict's Test	Barfoed's Test	Seliwanoff's Test	Iodine Test
#1 Glucose				
#2 Fructose				
#3 Sucrose				
#4 Lactose				
#5 Starch				
#6 Water				
#7 Unknown: ____________				
S1 Saliva/Starch				
S2 Saliva/Starch				

Other comments/observations:

TA Signature

Pre-Lab Assignment _________
Safety/Participation _________
Lab Write-Up _________

Ask your TA to review your work and sign your report. The TA will sign above once satisfied that the student has performed the entire procedure. The report will not be accepted or graded unless signed.

POST-LAB

Conclusions

1. Based on the data collected, determine the identity of the unknown carbohydrate. Include information about the functional group on your unknown as well as the number of carbons and possibility of your unknown being a disaccharide or polysaccharide. Give specific reasoning for this conclusion that is supported by the data.

2. What conclusions can be drawn about the action of saliva on starch? What do the tests indicate is the product of the reaction between starch and saliva? Give specific reasoning supported by data.

Questions

1. Why do ketoses need to isomerize (transform) into aldoses to react with the Benedict's reagent?

2. What enzyme in saliva is used to break down starch?

Synthesis of Organic Compounds
Aspirin and Wintergreen Oil

Learning Objectives

- Recognize esterification reactions.
- Understand and apply the laboratory techniques of suction filtration and recrystallization.
- Calculate theoretical and percent yield of chemical reactions.
- Using collected data, explain the results of an experiment in lab report form.

Additional Reading and References

- Timberlake: 7.4–7.8, 16.3

Safety Precautions and Hazards

- Chemical splash goggles and lab coat must be worn at all times.
- The product from this experiment will be collected for proper disposal.

PRE-LAB QUESTIONS

Name: _______________________________   Partners: _______________________________

TA: _______________________________   Section: _____________   Date: _______________

Answer the following questions using information obtained from lecture, the textbook, or lab manual. The responses will be checked at the start of your lab section for credit. Failure to complete may result in exclusion from participating in the lab.

1. Write the equation for the synthesis of aspirin to be used in this experiment.

2. Write the equation for the synthesis of wintergreen oil.

3. Fill in the blanks for this reaction table for the synthesis of aspirin, a common sight and pre-lab assignment for most chemists running organic reactions.

Equivalents	Compound	Moles	MW, d	Amount	Hazards
1.0 eq	Salicylic Acid	36.2 mmol	138.12 g/mol		Toxic
1.75 eq	Acetic Anhydride	63.5 mmol	102.09 g/mol 1.082 g/mL		
Catalytic	Sulfuric Acid	6.4 mmol	~18 mol/L	8 Drops	

4. Calculate the theoretical yield for aspirin based on the amount of reagents used above.

5. What is the proper procedure for checking the smell of a chemical or reaction (if needed)?

6. If time permits, melting points will be collected to verify the proper synthesis of aspirin. What is the melting point of aspirin? Salicylic acid?

BACKGROUND

Willow bark was known to the ancients to have fever-reducing (antipyretic) power. Modern investigators found that this was due to a natural organic compound in the bark, salicin. This bitter glucoside (a derivative of glucose) was first isolated in 1827. Hydrolysis of salicin yields saligenin and glucose (Figure 5.1):

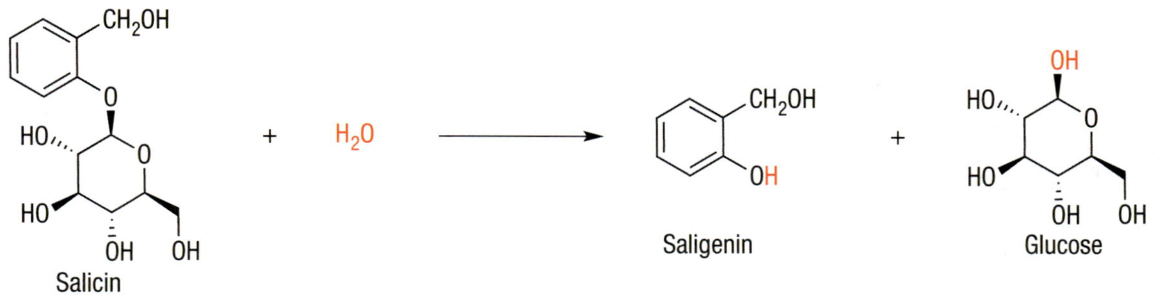

FIGURE 5.1 Hydrolysis of Salicin

Saligenin can be subsequently be oxidized to salicylaldehyde which, in turn, can be further oxidized to salicylic acid (Figure 5.2).

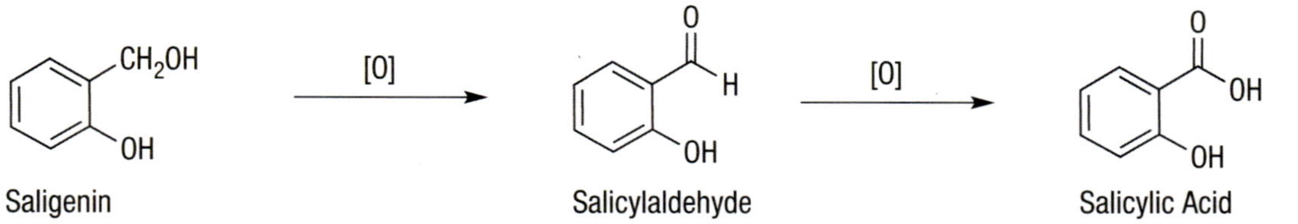

FIGURE 5.2 Oxidation of Saligenin to Form Salicylic Acid

The sodium salt of salicylic acid was initially prescribed for its antipyretic and analgesic effects, but the salt was too irritating to the stomach lining, so a phenyl ester version was developed as a replacement. While the esterified version passed successfully through the acidic stomach without irritation, it was hydrolyzed in the basic pH environment of the intestines, producing poisonous phenol. To avoid this problem, acetylsalicylic acid, where an ester is instead formed by the reaction of the alcohol group of salicylic acid with acetic acid, is now used. Acetylsalicylic acid, better known by the common name aspirin, gets hydrolyzed in the intestines to release salicylic acid (the actual analgesic substance) and acetate. The typical aspirin tablet is five grains (~0.324 grams), but in larger quantities becomes toxic.

Isolating aspirin from living plants would be an extremely expensive process, so it is now predominately synthesized using petroleum byproducts like benzene which are cheaply obtained. The total synthesis of aspirin from benzene requires multiple steps, many of which are beyond the scope of this course. In this experiment, aspirin will be simply synthesized from salicylic acid using acetic anhydride (Figure 5.3).

An anhydride is a type of carboxylic acid derivative formed from the dehydration of two carboxylic acid molecules. Anhydrides will react with water to form the corresponding carboxylic acid, or with an alcohol to form a carboxylic acid and an ester:

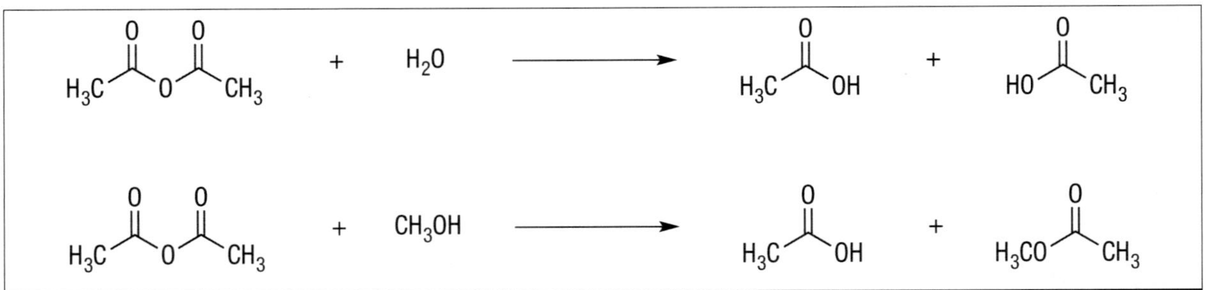

FIGURE 5.3 Common Reactions of Acetic Anhydride with Water and Methanol

Acetic anhydride is used in this instance to create an ester because it is more reactive than acetic acid and makes the reaction occur more quickly. In our experiment, sulfuric acid will also be used as a catalyst to help speed up the reaction (Figure 5.4).

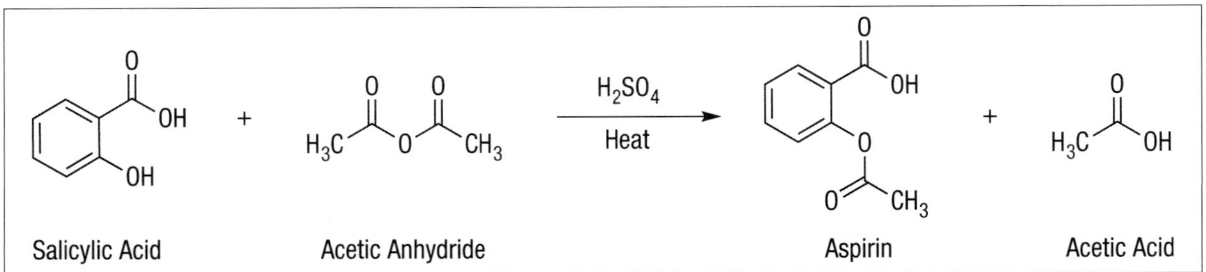

FIGURE 5.4 Reaction Scheme for the Synthesis of Aspirin

The aspirin product is nearly insoluble in water as in transforming the salicylic acid the opportunities for hydrogen-bonding have significantly decreased. This makes it easy to separate the aspirin from the rest of the reaction mixture using filtration as everything else will be water-soluble.

Another reaction of salicylic acid will also be investigated in this experiment. Oil of wintergreen (methyl salicylate) can be easily prepared through an esterification reaction with salicylic acid. Wintergreen oil is commonly used as a flavoring agent and in rubbing ointments. Methyl salicylate is made by reacting methanol with the acid functional group of salicylic acid as shown in Figure 5.5. A successful transformation using this reaction will easily be noted by the natural fragrance of the ester.

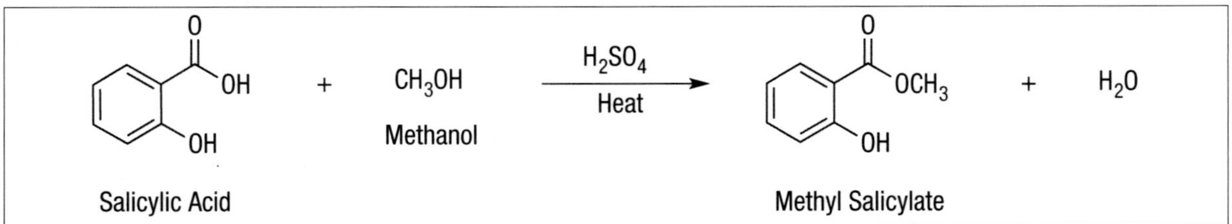

FIGURE 5.5 Reaction Scheme for Synthesis of Oil of Wintergreen

PROCEDURE

At several points you will use a water bath for heating. Use a few boiling chips in the water bath to avoid "bumping."

A. Preparation of Aspirin

1. Into a weigh boat, weigh out about 5.00 g of salicylic acid to the nearest 0.01 g and transfer it into a 125-mL Erlenmeyer flask.

2. Add 6.0 mL acetic anhydride to the salicylic acid.

Acetic anhydride acid is a flammable liquid and vapor. It is harmful if swallowed and causes severe skin burns and eye damage. Toxic if inhaled. May cause respiratory irritation.

3. Carefully, add eight drops of concentrated sulfuric acid, swirling the flask gently.

Concentrated H_2SO_4 is extremely corrosive to skin and clothing and reacts violently with water.

4. Set up a water bath by clamping the flask to a ring stand and placing a 600-mL beaker under the flask.

5. Fill the beaker about two-thirds full of water until the bottom of the flask is about one half inch into the water.

6. Heat the system on a hot plate for about 15 minutes after the water begins to boil gently. If the solid does not dissolve, heat for an additional 10 minutes, stirring as necessary.

7. Remove the flask and add 25 mL of ice water to the flask to decompose any unreacted anhydride.

8. Set the flask in a beaker of ice until crystallization is complete.

9. Separate the crystals from the liquid by suction filtration. Your TA will show the proper arrangement for this.

10. Wash the crystals well with water, breaking up any large chunks.

11. Leave the crude crystals to dry on the suction filter while you perform the experiment in Part B.

12. Measure the crude yield. Make additional observations about the nature of the crystals.

13. Recrystallize the crude aspirin by dissolving the crystals in a small amount of ethanol in a small beaker. Heat the solution to boiling. All the crystals should eventually dissolve when boiling is reached (if sufficient ethanol was used). Carefully remove the beaker from the hot plate and allow to cool to room temperature. Promote further crystallization by cooling in an ice bath. (Should this recrystallization prove tricky, ask your TA for advice on how to proceed. Recrystallizations are more of an "art" than a "science.")

14. Use the suction filtration apparatus and a clean piece of filter paper to filter the crystals again. Use some water to wash if necessary.

15. Allow the crystals to dry, breaking up any lumps with a spatula.

16. Weigh the aspirin again to get the purified yield.

17. Measure the melting point of the aspirin and compare with that of salicylic acid and pure acetylsalicylic acid. (Ask your TA for specific directions on how to prepare the samples and how to use the melting point apparatus.)

18. Place any leftover product into the designated waste container.

B. Preparation of Oil of Wintergreen

19. In a small beaker (or a test tube), add ~2 mL methanol and a tiny amount of salicylic acid (about the size of a match head).

Methanol is a highly flammable liquid and vapor. It is toxic if swallowed, inhaled, or comes in contact with skin.

20. To this mixture add 10 drops sulfuric acid.

21. Carefully warm the mixture in a water bath.

22. After a few minutes in the bath, carefully smell the mixture by gently waving your hand over the beaker towards your nose.

23. Note the change in odor going from reactants to products.

24. Discard all waste in the appropriate waste container.

DATA SHEET

Name: _______________________________   Partners: _______________________________

TA: _________________________________   Section: ____________   Date: ________________

Notes and Observations

TA Signature	
Pre-Lab Assignment _________ Safety/Participation _________ Lab Write-Up _________	Ask your TA to review your work and sign your report. The TA will sign above once satisfied that the student has performed the entire procedure. The report will not be accepted or graded unless signed.

POST-LAB

The post-lab assignment for this experiment is a lab report. Broad directions are provided in Appendix B of this lab manual. Your TA should have some additional suggestions or guidelines that specifically pertain to this experiment. As they are the ones who will be grading the report, it is important to follow their specific requirements.

What follows here are some ideas to consider as you are writing your reports:

- What happened to any unreacted acetic anhydride when the reaction mixture was added to the ice water? Write an equation for it.

- Give some reasons why the actual yield is less than the theoretical yield.

- Can the percent yield ever be (or appear to be) more than 100%? Explain.

- Are we able to definitively conclude that we successfully synthesized aspirin? What other pieces of data would you like to collect to be sure?

- What is the relationship between the synthesis of aspirin and the synthesis of wintergreen oil? Why are these performed together?

Fats, Oils, Soaps, and Detergents

Learning Objectives

- Recognize the difference between fats and oils.
- Understand saponification reactions involving triacylglycerols and the products that result.
- Connect the topics of triacylglycerols and hydrolysis to soaps and detergents used in daily life.
- Write chemical reactions involving carboxylate salts.
- Describe how a soap/detergent is able to facilitate solubility in water of non-polar molecules (grease).

Additional Reading and References

- Timberlake: 17.1, 17.2, 17.3, 17.4

Safety Precautions and Hazards

- Chemical splash goggles and lab coat must be worn at all times.
- Nitrile gloves must be worn when handling chemicals.
- Pay special attention to cleanliness throughout this procedure.

PRE-LAB QUESTIONS

Name: _______________________________   Partners: _______________________________

TA: _______________________________   Section: _____________   Date: _______________

Answer the following questions using information obtained from lecture, the textbook, or lab manual. The responses will be checked at the start of your lab section for credit. Failure to complete may result in exclusion from participating in the lab.

1. Draw two triacylglycerol molecules below, one that would be considered a *fat* and one that would be considered an *oil.*

fat

oil

2. The lab manual describes two reasons why soaps or detergents are basic. List those reasons below:

  a.

  b.

3. Describe how a positive result for the Baeyer Test will look. What is being oxidized? What is being reduced?

4. Which kind of triacylglycerol should test positive for this test? Why?

BACKGROUND

Properties of Triacylglycerols

Fats and oils of animal or vegetable origin are triacylglycerols, or triesters of glycerol with long-chain fatty acids (Figure 6.1). The length of the fatty acids that comprise the fats or oils may vary in length and in degree of unsaturation (presence of carbon–carbon multiple bonds). The most common fatty acids which occur in fats and oils have twelve to eighteen carbon atoms. If the long hydrocarbon chains are saturated or nearly so, the triacylglycerols are solids at room temperature and are commonly called **fats.** Double bonds in the hydrocarbon chains will lower the melting point of triacylglycerols due to disruptions in the intermolecular dispersion forces; thus highly unsaturated triacylglycerols tend to be liquids and are referred to as **oils.** Vegetable oils, such as cottonseed or peanut oil, can be converted to solid fats by hydrogenating some of the double bonds and fully saturating the hydrocarbon chains.

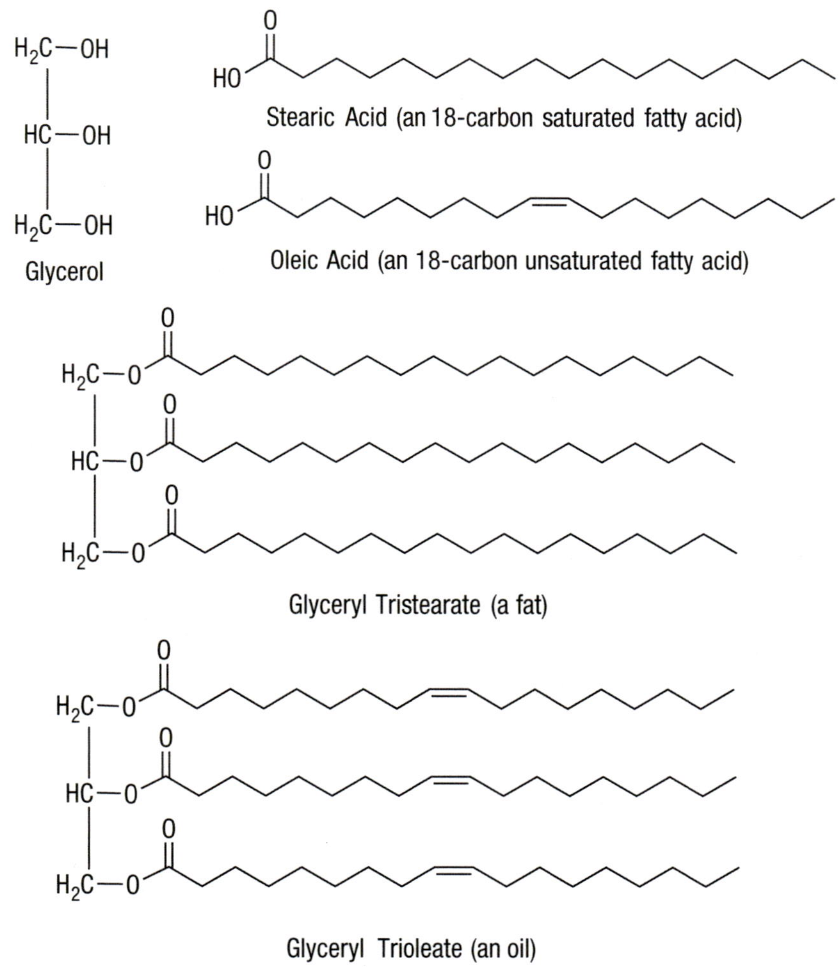

FIGURE 6.1 Structural Representations of Glycerol, Fatty Acids, and Common Triacylglycerols

Preparation of Soap

The reaction of a fat or an oil with aqueous base (Figure 6.2) is known as base hydrolysis or saponification (from the Latin "sapo" meaning soap). When saponified, the fat or oil is converted to glycerol and the salts of the fatty acids. These carboxylate salts are called **soaps.** This kind of reaction is the basis of the soap industry and is also a commercial source of glycerol, which itself has uses as an antifreeze, tobacco moistening agent, or even as a reactant in manufacturing nitroglycerine.

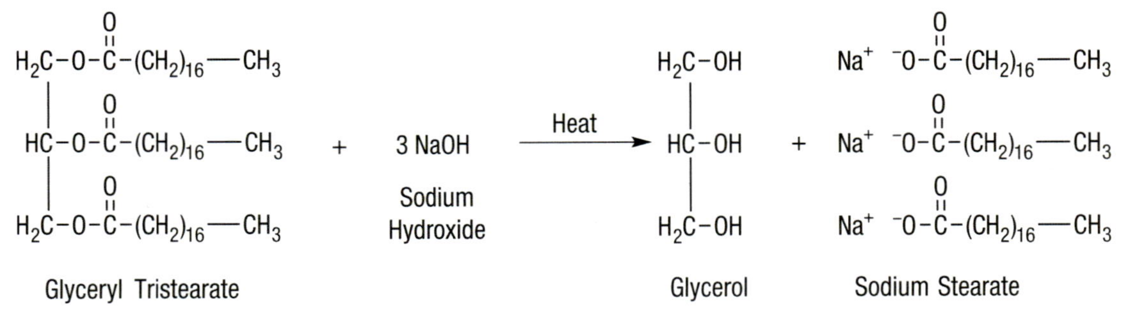

$$H_2C-O-\overset{O}{\overset{\|}{C}}-(CH_2)_{16}-CH_3$$
$$HC-O-\overset{O}{\overset{\|}{C}}-(CH_2)_{16}-CH_3 \quad + \quad 3\ NaOH \quad \xrightarrow{Heat} \quad$$
$$H_2C-O-\overset{O}{\overset{\|}{C}}-(CH_2)_{16}-CH_3$$

Glyceryl Tristearate   Sodium Hydroxide

$$H_2C-OH$$
$$HC-OH \quad + \quad Na^+\ {}^-O-\overset{O}{\overset{\|}{C}}-(CH_2)_{16}-CH_3$$
$$H_2C-OH$$

Glycerol   Sodium Stearate

FIGURE 6.2 Saponification (Base Hydrolysis) of Glyceryl Tristearate

Properties of Soaps and Detergents

Soap and detergents have many common properties due to the polar, charged regions within their structures. Although the sodium salts of long-chain fatty acids are water-soluble, the acids themselves are not. Thus acidification of a soap solution causes the fatty acid to precipitate. Stearic acid ($C_{17}H_{35}COOH$) may be prepared this way and is commonly mixed with paraffin for use in the manufacture of candles. Some of the properties of both soaps and detergents that will be explored are as follows:

Alkalinity

The alkalinity of a soap or detergent comes from two sources. The first is the residual base from the saponification of the fat or oil. The second source is from the ability of the carboxylate salts, which are weak bases, to react with water to produce the parent carboxylic acid and a hydroxide ion.

Detergent Properties

The cleansing action of soaps and detergents arise from the interaction of the hydrocarbon chain of the fatty acid ("tail") with greasy dirt particles, and carboxylate salt ("head") with water. The greasy dirt is broken up by the soap or detergent, and encapsulated inside a hollow ball formed by the soap or detergent. This hollow ball, more appropriately called a **micelle,** can be flushed away from the surface being cleaned with water. The process by which the dirt is broken up and encapsulated is known as **emulsification.**

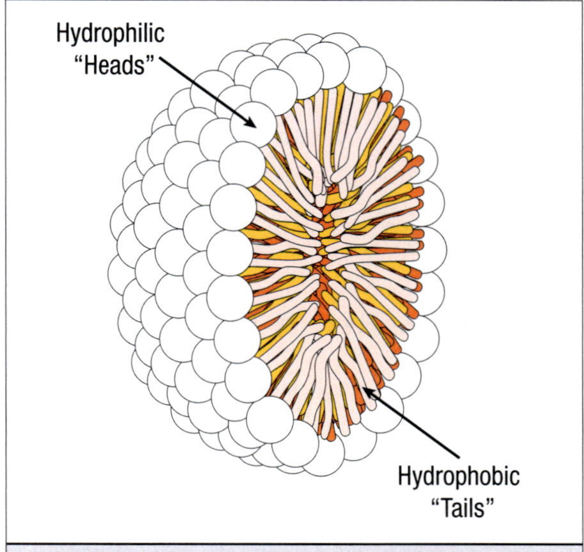

FIGURE 6.3 Micelle (shown as a cross-section of the spherical structure)

Effect of Hard Water on Detergency

Although the sodium and potassium salts of ordinary soaps are water-soluble, their calcium, magnesium, and iron salts are not. The latter ions are present in "hard" water, which has not been deionized to the same extent as the "soft" water, and consequently, they form precipitates with ordinary soaps. This precipitation removes the soap from solution and decreases its effectiveness as a detergent. Most commercial detergents avoid this problem by using functional groups other than carboxylates such as aryl (aromatic) sulfonates or alkyl sulfates, whose calcium salts are water-soluble. These have the virtue of functioning almost as well in "hard" as in "soft" water.

FIGURE 6.4 Structure of Common Commercial Detergent

PROCEDURE

In this experiment, the chemical and physical properties of various kinds of triacylglycerol derivatives will be examined. Comparisons will also be made to detergents which possess similar physical properties to the naturally occurring lipids.

Start the preparation of soap (Part A); perform Part B while this reaction is taking place.

A. Preparation of Soap

1. Heat a large beaker partially filled with water to ~80–90°C (just below boiling).

2. Measure ~10 g lard into a small, tared beaker or Erlenmeyer flask.

3. In a separate beaker or flask, prepare a sodium hydroxide solution by dissolving 6 g NaOH in 25 mL water.

Sodium hydroxide, also known as caustic soda, causes severe skin burns and eye damage.

4. To the flask containing the lard, add ~15 mL ethanol and your prepared sodium hydroxide solution.

Ethanol is highly flammable.

5. Using a ring stand and clamp so it does not tip over, place the lard flask in the larger beaker of hot water.

6. Continue heating for approximately one hour maintaining the temperature of water bath between 80–90°C and stirring the mixture regularly. If a significant amount of the water and alcohol should evaporate such that the contents of the flask become almost solid, add a little distilled water. (**Note:** Do not stir the mixture using the thermometer. Use a glass stir rod as needed for this.)

While the saponification process occurs, examine the chemical and/or physical properties of various triacylglycerols using the subsequent steps:

B. Properties of Triacylglycerols

7. Place ten drops (~0.5 mL) cottonseed oil (or vegetable oil) into each of three test tubes.

8. Label one of the test tubes "water," another "ethanol," and the last one "hexanes." Add 1 mL of the respective solvent to the tubes.

Ethanol and hexane are both highly flammable.

9. Shake each tube, and note whether the oil dissolves. If any of the solvents should fail to dissolve the oil, add another 5 mL of that solvent and again shake vigorously to see if the oil will dissolve.

10. Record your observations as to the solubility of cottonseed oil in each of these solvent.

11. In another test tube, combine one pipette full of cottonseed oil and one pipette full of hexanes, agitating to combine.

12. In a second new test tube, combine one pipette of hexanes and one pipette full of a melted cooking fat ("Crisco" or margarine should be available). The fat may be melted by placing it on a watch glass and warming the watch glass on a hot water bath. Again, agitate to combine.

13. Add five drops of the Baeyer ($KMnO_4$) solution to both test tubes. Observe and record the results.

The Baeyer Test utilizes the oxidizing power of the permanganate ion (MnO_4^-). When reacted with a molecule containing a carbon–carbon double or triple bond, the purple color changes to brown as the bond is oxidized. When it is added to saturated hydrocarbons, the purple color of the permanganate ion is unchanged.

Continue to monitor the progress of the saponification of lard, but using some previously prepared soap move on to Part C.

C. Comparison of Soap and Detergents

Each of the following tests will be performed on previously prepared soap and a synthetic detergent.

14. Dissolve about 2 g of soap in 100 mL distilled water. You may need to gently heat the solution to promote full dissolution.

15. Dissolve about 2 g of a commercial detergent in 100 mL of distilled water. You may need to gently heat the solution to promote full dissolution.

Alkalinity

16. Using test tubes, test a sample of each solution using both pH paper and a few drops of phenolphthalein solution (which turns pink above pH 8–9).

Detergent Properties

17. Place four drops of mineral oil into each of three separate test tubes.

18. Add 5 mL distilled water to one tube, 5 mL of your soap solution to another, and 5 mL of detergent solution to the third tube.

19. Shake each tube for a minute, then observe how well the oil is emulsified.

Effect of Hard Water on Detergency

20. Place 5 mL of your soap solution in each of three test tubes.

21. To each tube, add 2 mL of a 1% solution of calcium, magnesium, or ferric chloride, and note whether a precipitate forms.

22. Test each of these solutions for its emulsifying power, by adding four drops of mineral oil to each tube and shaking for a minute.

23. Repeat, using the detergent solution, and compare the results.

A. Preparation of Soap (continued)

24. After the hour of heating has elapsed, carefully add 200 mL saturated salt solution.

    Adding the salt to the soap mixture precipitates the soap due to the common ion effect. The soap is a sodium salt and when excess sodium ion is added it causes the soap to fall out of solution.

25. Cool the mixture and filter it through a double thickness of cheesecloth.

26. Rinse the soap on the filter with 50 mL of cold water.

27. Press the soap into a small evaporating dish which will serve as a mold.

28. At the end of class, place your synthesized soap into the appropriately labeled box located in one of the hoods.

DATA SHEET

Name: _________________________________   Partners: _________________________________

TA: _________________________________   Section: _____________   Date: _____________

Properties of Triacylglycerols

1. Solubility of Cottonseed/Vegetable Oil

Solvent	Solubility Observations
Water	
Ethanol	
Hexane	

2. Unsaturation

Substance	Observations
Cottonseed/Vegetable Oil	
Fat	

Properties of Soaps and Detergents

Test	Observations of Pre-Prepared Soap	Observations of Detergent
Alkalinity (pH paper)		
Alkalinity (phenolphthalein)		
Detergency		
Calcium Ion		
Magnesium Ion		
Ferric Ion		

TA Signature

Pre-Lab Assignment ________

Safety/Participation ________

Lab Write-Up ________

Ask your TA to review your work and sign your report. The TA will sign above once satisfied that the student has performed the entire procedure. The report will not be accepted or graded unless signed.

POST-LAB

Questions

1. Which of the solvents tested (water, ethanol, or hexanes) would you select to remove a gravy stain from your clothes? What factors governed your choice?

2. Write the equation for the reaction of glyceryl trioleate with bromine.

3. How could you distinguish a solution of a fat from a solution of a vegetable oil?

4. Why is the hydrolysis of a fat called "saponification"?

5. Explain why the hydrolyzed fat mixture was poured into a concentrated salt solution to precipitate the soap.

6. What impurities were removed from the soap by washing it with cold water?

7. Assuming a soap to be pure sodium stearate, give an equation

  a. which shows why its solutions are alkaline.

  b. for its reaction with Mg^{2+} in hard water.

  c. for its reaction with hydrochloric acid.

Analysis of Organic Compounds

Learning Objectives

- Develop skills and procedures useful in the identification of simple organic compounds.
- Reinforce a knowledge of nomenclature and physical and chemical properties of simple organic compounds.
- Use qualitative observations to classify and identify organic compound based on functional group.
- Design a set of experimental steps using a flow chart.

Additional Reading and References

- Timberlake: 12.4, 13.3, 14.2, 16.2, 18.2

Safety Precautions and Hazards

- Chemical splash goggles and lab coat must be worn at all times.
- Nitrile gloves must be worn when handling chemicals.
- No reagents from this experiment can go down the drain.
- There are multiple chemical hazards in this experiment, so pay attention to the safety notes throughout the procedure.

PRE-LAB QUESTIONS

Name: ____ Partners: ____

TA: ____ Section: ______________ Date: ________________

Answer the following questions using information obtained from lecture, the textbook, or lab manual. The responses will be checked at the start of your lab section for credit. Failure to complete may result in exclusion from participating in the lab.

1. What is the difference between qualitative observations and quantitative observations?

2. Using your own words, summarize the expected results for each of the tests that will be used to identify functional groups in this experiment.

3. You are presented with the following compound: 2–propanol. Describe below the series of experiments that would be necessary to perform (and what observations you would expect) to conclusively verify qualitatively that it contains an alcohol.

BACKGROUND

Qualitative analysis differs from quantitative analysis in that no measurements of the substances are recorded, only observations of their properties or reactions. The techniques to be used in this experiment will give valuable information about the functional groups present in each compound, but give no specific information about the compound being tested. Scientists often make qualitative observations to determine what set of quantitative experiments might be appropriate to proceed with in understanding a particular synthesized compound.

The laboratory exercise to be performed here will permit you to tangibly observe some of the differences in chemical and physical properties of functional groups first-hand. The major areas that will be explored are: solubility, acid-base characteristics, and chemical reactivity. A series of simple, qualitative experiments may be done in series with one another to conclusively identify the presence of a particular functional group. The experiments to be utilized are described below, classified by broad chemical or physical property.

Solubility

Water soluble compounds are usually low molecular weight and have functional groups which may hydrogen-bond to the water molecules. Water insoluble compounds are likely to have higher molecular weights and be carbon-rich structures that generally cannot hydrogen-bond. For many functional groups, general rules of thumb for solubility are described in the Additional Reading.

Acid-Base Characteristics

Acid and bases are commonly defined using the Brønsted-Lowry model. Acids are those molecules that act as proton donors, while bases act as proton acceptors. The ability of a functional group to donate or accept a proton makes for very unique properties, both chemical and physical. There are two simple functional groups that possess acid-base characteristics: carboxylic acids and amines. For carboxylic acids, the hydrogen that is attached to the carboxylate group is easily removed when the group is either dissolved in water or a base is added to mix with it. For amines, the nitrogen is easily protonated when placed in water or acidic conditions. All other functional groups that have been discussed are not effected by different acid-base (or pH) conditions. By using a pH indicator or pH paper, the change in pH of a solution may be observed. If a carboxylic acid is present, the solution will be become more acidic. If an amine is present, the solution will become more basic.

Chemical Reactivity

Chemical reactivity is a great way to distinguish one functional group from another. Some molecules are easily oxidized, others are not. Some functional groups react distinctly with specific reagents to form products that may precipitate out of solution or change colors in unique ways. By choosing the right experiments, particular organic functional groups may be distinguished without making any quantitative measurements.

As was previously observed in Experiment 6 during the oxidation of unsaturated fatty acids, an oxidant like Baeyer's reagent, which utilizes the oxidizing power of the permanganate ion (MnO_4^-), can be used to mediate the transformation of a molecule containing a carbon-carbon double into an alcohol (Figure 7.1). The Baeyer's reagent changes from a purple color to brown as the double bond is oxidized. However, when it is added to a solution of a saturated hydrocarbon, the purple color of the permanganate ion remains unchanged as there is no group to oxidize. Most common oxidizing agents are not particularly selective as to the functionality that they will oxidize, so care must be taken when analyzing the results as reactions with other electron-rich functional groups such as aldehydes may also occur.

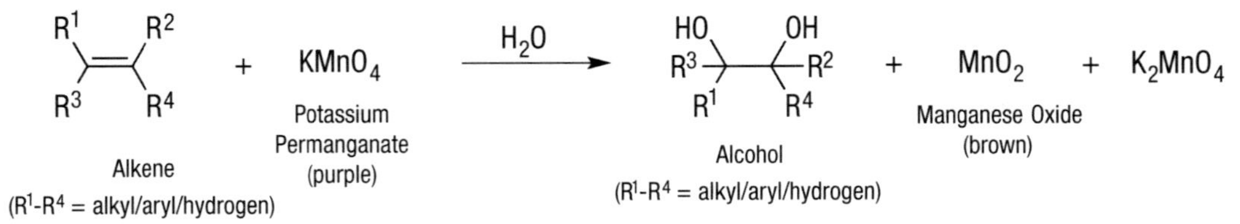

FIGURE 7.1 Oxidation of Double-Bonds Using Baeyer's Reagent

In Experiment 5 another kind of oxidation reaction was observed using Benedict's reagent. This involved oxidizing aldehyde-containing carbohydrates to make carboxylic acid-containing groups called alditols. Benedict's reagent can be generally used to oxidize other non-carbohydrate containing aldehydes. During the reaction, copper ions in Benedict's reagent, which are blue in color, are reduced as the aldehyde is being oxidized forming a copper oxide precipitate and the new carboxylic acid (Figure 7.2). The color of this copper precipitate can range from green to yellow to brick red. If no reaction occurs, the compound being examined does not contain an aldehyde functional group.

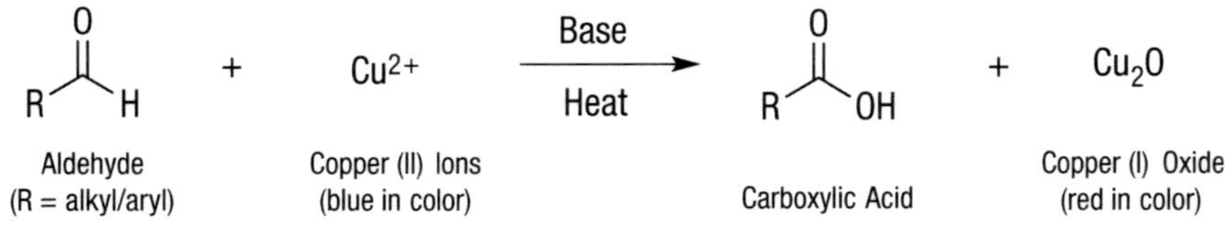

FIGURE 7.2 Oxidation of Aldehydes Using Benedict's Reagent

Another functional group, carboxylic acids, while unable to be oxidized, have other distinct properties to exploit. Due to their acidic characteristics, carboxylic acids will react chemically when mixed with bases. One such reaction happens when organic acids are added to a sodium bicarbonate solution. Neutralization of the acid occurs, forming a carboxylate salt and, as the byproduct of base, carbonic acid. The carbonic acid immediately decomposes to form carbon dioxide and water. The carbon dioxide can be seen as bubbles escaping the solution and provides visual evidence of a successful reaction (Figure 7.3).

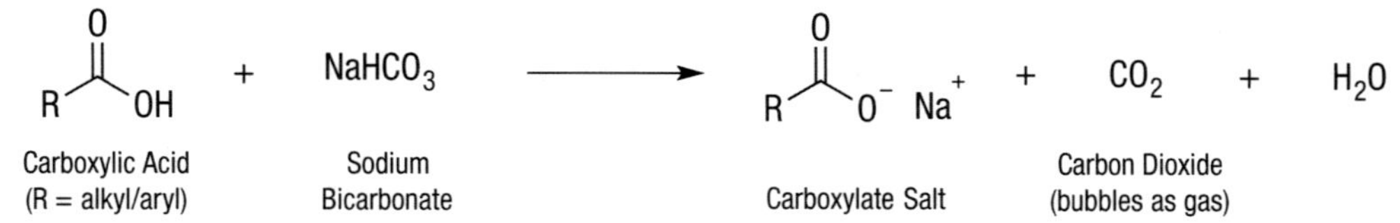

$$R-C(=O)-OH \;+\; NaHCO_3 \;\longrightarrow\; R-C(=O)-O^- \; Na^+ \;+\; CO_2 \;+\; H_2O$$

Carboxylic Acid (R = alkyl/aryl) | Sodium Bicarbonate → Carboxylate Salt + Carbon Dioxide (bubbles as gas)

FIGURE 7.3 Acid-Base Reactions of Carboxylic Acids Using Sodium Bicarbonate

Aside carbonyl compounds, hydroxyl-group containing functionality like alcohols and phenols may be distinguished by using their unique able to complex with ceric nitrate. In this reaction, some of the groups attached to the Ce-metal (ligands) are swapped for either alcohols or phenols (Figure 7.4). This changing of ligands results in a color change from yellow to red in the presence of alcohols and from yellow to blackish-brown in the presence of phenols.

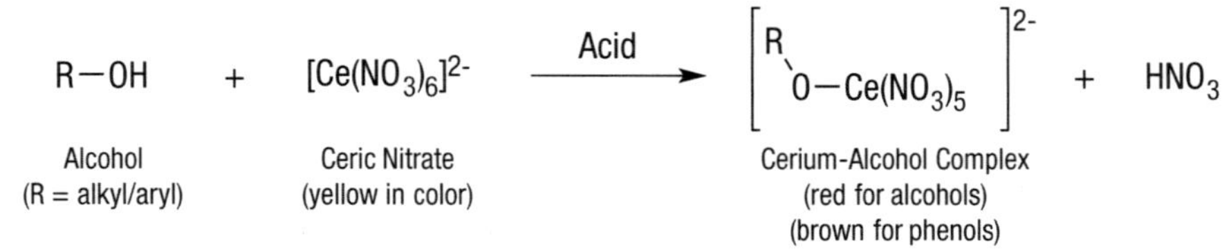

$$R-OH \;+\; [Ce(NO_3)_6]^{2-} \;\xrightarrow{\text{Acid}}\; \left[\begin{array}{c} R \\ O-Ce(NO_3)_5 \end{array} \right]^{2-} \;+\; HNO_3$$

Alcohol (R = alkyl/aryl) | Ceric Nitrate (yellow in color) | Cerium-Alcohol Complex (red for alcohols) (brown for phenols)

FIGURE 7.4 Metal Complexation of Alcohols and Phenols with Ceric Nitrate

Phenol and alcohols may be further distinguished by using their differing ability to complex with ferric chloride (Figure 7.5). When phenols are mixed with a ferric chloride solution, a color change from yellow to purple (or red, blue, green depending upon what other groups are attached to the phenol) may be observed. Alcohols will manifest no change in color of the solution.

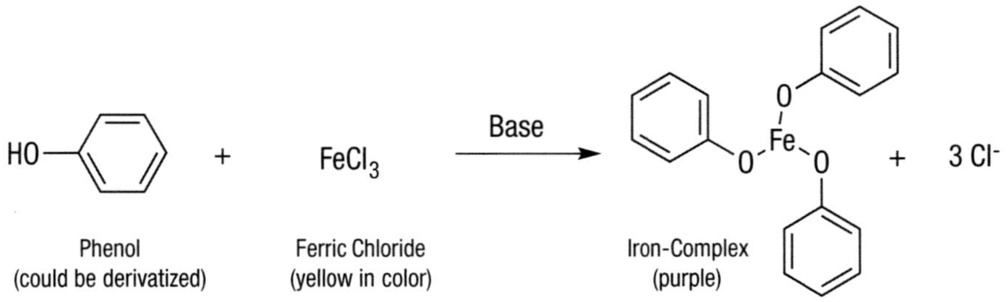

Phenol (could be derivatized) | Ferric Chloride (yellow in color) → Iron-Complex (purple) + $3\,Cl^-$

$$HO-C_6H_5 \;+\; FeCl_3 \;\xrightarrow{\text{Base}}\; \text{Fe(OC}_6\text{H}_5)_3 \;+\; 3\,Cl^-$$

FIGURE 7.5 Metal Complexation of Phenols with Ferric Chloride

Aldehydes and ketone are easily differentiated from other functional groups via the reaction with Brady's reagent (Figure 7.6). Most aldehydes and ketones produce an insoluble precipitate called a dinitrophenylhydrazone upon reaction with the active component, 2,4–dinitrophenyhydrazine (2,4–DNP). The insoluble precipitate may at first appear oily but often becomes crystalline upon standing.

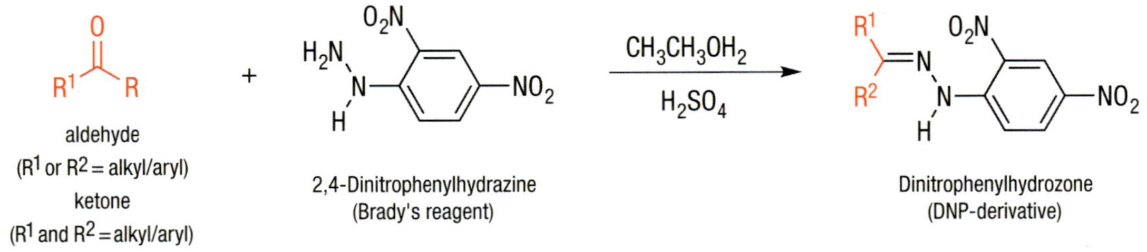

FIGURE 7.6 Aldehydes and Ketones Reaction with Brady's Reagent

Finally, amines may be distinguished from all other functional groups through the use of the Hinsberg reaction. Primary amines will react with benzenesulfonyl chloride to give an amide that is soluble in NaOH solution (Figure 7.7). Secondary amines when reacted with benzenesulfonyl chloride on the other hand will produce an amide that is insoluble in NaOH. Tertiary amines do not react with benzenesulfonyl chloride.

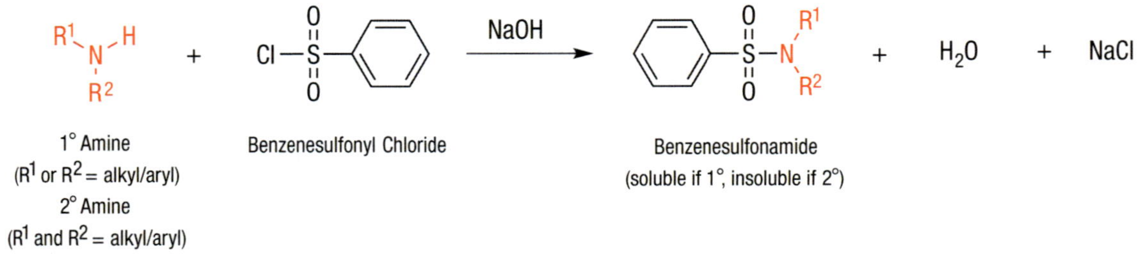

FIGURE 7.7 Hinsberg Reaction with Amines

When used in tandem, all of these properties and reactions can be used to differentiate compounds containing indeterminate functional groups. A flow chart (Figure 7.8) has been generated to help organize these reactions. Using this chart, it should be possible to qualitatively determine the identity of some simple organic molecules.

PROCEDURE

For this experiment there is not a specific proscribed set of steps you need to follow. The provided flow chart (Figure 7.8) should be used to guide you on the series of qualitative experiments that might be needed to identify a particular unknown compound. There are 8 unknowns that are labeled A through H that need to be identified as to their functional group. What follows here in the procedure section is guidance in performing each named experiment depicted on the chart. Please note that is not necessary to perform every test. You should only perform those required to identify the functionality in your unknowns. **All** materials in this lab need to be placed into a waste container. Nothing should be put down the drains.

A. Solubility

The solubility of a compound gives clues to its structure. Distilled water will be used as the solvent for these tests.

Test for water solubility by adding ~10 drops of the compound to a test tube containing one pipette of water. A compound is considered soluble if it completely dissolves in the water. When judging the solubility of two clear liquids, look for the formation of layers or the presence of undissolved droplets to indicate an insoluble compound.

B. Acid-Base Properties

Based on the flow chart, these tests should only be performed with water-soluble compounds. Before performing any acid-base experiments, first test the pH of the distilled water for use as a comparison. In a test tube, add three drops of universal indicator to one pipette full of distilled water. Record the color. If available, also test the pH of water using pH paper. To the test tubes of the water soluble compounds, add three drops of universal indicator. Based on the color change you should be able to classify the unknown compound as:

- **Acidic:** If the color changes to red, the compound is strongly acidic. If the color is yellow, the compound is slightly acidic.
- **Neutral:** If the color is similar to the color of the distilled water used as a comparison the compound is neither acidic nor basic.
- **Basic:** If the color is dark blue or purple, the compound is basic.

Confirm your results by using pH paper to get a rough measure of the pH. If the results are inconsistent, the experiment should be repeated in consultation with your TA.

C. Qualitative Tests for Functional Groups

Before proceeding with any of these other tests, obtain clean test tubes, dirty ones may lead to false positives.

Unsaturation Test Using Baeyer's Reagent (KMnO$_4$)

To perform the test, combine one pipette full of the unknown and one pipette full of water in a clean test tube then agitate to mix thoroughly. Add 5 drops of the Baeyer (KMnO$_4$) solution to the test tube. Observe and record the results.

Carboxylic Acids Test Using Sodium Bicarbonate

To perform the test, use the tip of the spatula to add a small amount of sodium bicarbonate to a test tube. Add 10 drops of the compound to be tested and a few drops of distilled water. Observe and record the results.

Alcohols and Phenols Test Using Ceric Nitrate

To perform the test on water-soluble compounds, place half a pipette of ceric ammonium nitrate reagent in a test tube. Add ~5 drops of the compound to be tested. Observe and record the results.

Ceric ammonium nitrate is an oxidizer that can cause skin and eye irritation.

Phenols Test Using Ferric Chloride

Shake the ferric chloride solution before use. To perform the test, place half of a pipette of the compound to be tested into a test tube. Add three drops of the ferric chloride solution to the test tube. Observe and record the results.

Ferric chloride is corrosive and causes serious skin burns and eye damage.

Aldehydes and Ketones Test Using Brady's Reagent (2,4–DNP)

To perform the test, combine 2–3 drops of the compound to be tested with half a pipette of ethanol. Add to this half a pipette of 2,4–dinitrophenylhydrazine reagent. Agitate the test tube to mix. If no precipitate immediately forms, let the solution stand for ~15 minutes before making final observations.

Brady's reagent contains 2,4-dinitrophenylhydrazine that can cause damage to organs and irritate skin. This solution also contains sulfuric acid, which is corrosive and ethanol, which is highly flammable.

Aldehydes Test Using Benedict's Reagent

To perform the test, mix half a pipette of the compound to be tested with a full pipette of water. Add a full pipette of Benedict's reagent. Agitate to mix thoroughly and place in a hot water bath. The reaction may take a few seconds to begin. Observe and record the results.

Benedict's reagent contains copper sulfate, which can be an irritant.

Amines Test Using the Hinsberg Reaction

This test MUST be performed in the hood.

Place a full pipette of 3 M NaOH solution in a test tube and add ~6 drops of benzenesul-fonyl chloride and ~4 drops of the compound to be tested. The formation of a precipitate indicates a reaction; agitate the test tube to see if the precipitate dissolves in the NaOH solution. Observe and record the results.

Benzenesulfonyl chloride causes severe skin burns and eye damage and may cause an allergic skin reaction.

Isobutylamine is a highly flammable liquid and vapor that is toxic if swallowed and causes severe skin burns and eye damage.

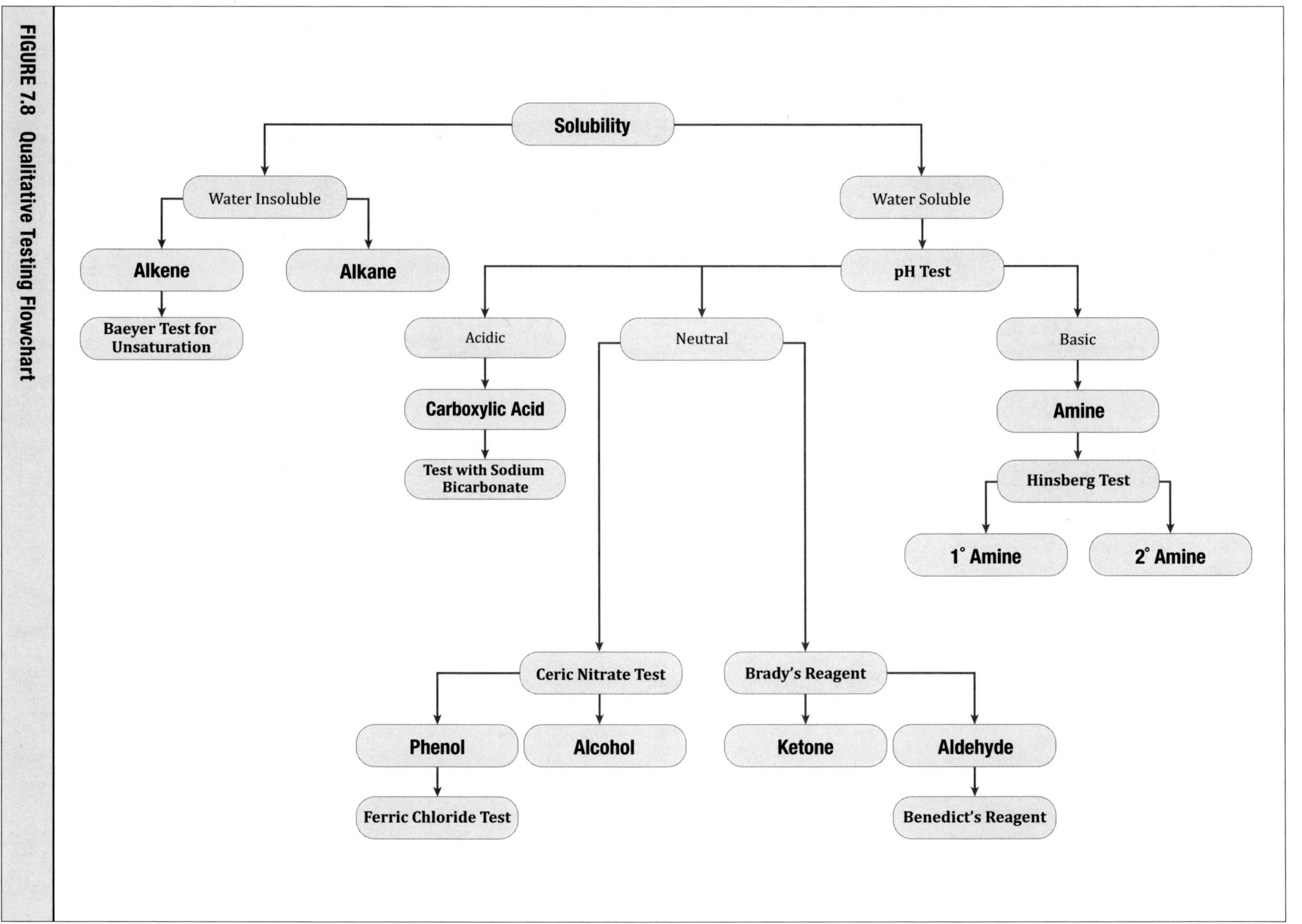

FIGURE 7.8   Qualitative Testing Flowchart

DATA SHEET

Name: ___________________________   Partners: ___________________________

TA: ___________________________   Section: ___________   Date: ___________

Data Table

Unknown	Experimental Design and Observations
A	
B	
C	
D	

Unknown	Experimental Design and Observations
E	
F	
G	
H	

TA Signature

Pre-Lab Assignment ________
Safety/Participation ________
Lab Write-Up ________

Ask your TA to review your work and sign your report. The TA will sign above once satisfied that the student has performed the entire procedure. The report will not be accepted or graded unless signed.

POST-LAB

For this lab, type up your responses (or answer on separate paper).

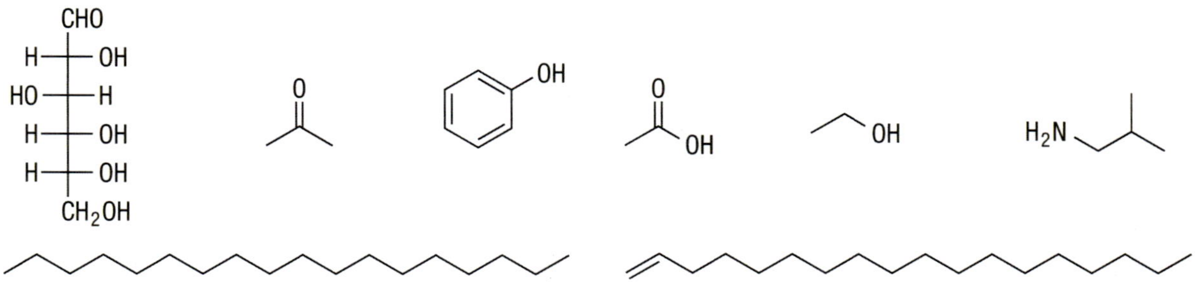

1. Give the names for each of the potential unknown structures above using IUPAC nomenclature. (**Note:** You may need to do some external research to give the names for the large
   hydrocarbon molecules.) For each unknown, write a paragraph that describes the series
   of experiments you used to determine its identity. Include any observations and how you
   were able to eliminate other similar structures from consideration.

2. Which functional groups were soluble in water? Which groups were insoluble in water?
   Relate the solubility observations to hydrogen bonding ability of the molecules.

Paper Chromatography

Learning Objectives

- Understand principles associated with chromatography including mobile phase and stationary phase.
- Categorize amino acids as neutral non-polar, neutral polar, acidic, and basic.
- Recognize how changing the mobile phase polarity effects the retention factor.
- Calculate retention factors for the various spots on a chromatogram.
- Make connections between classroom content and laboratory content and expound upon them in writing.

Additional Reading and References

- Timberlake: 19.1, 19.2
- McCaldin, D.J. "The Chemistry of Ninhydrin." *Chem. Rev.* **1960,** *60,* 39–51.

Safety Precautions and Hazards

- Chemical splash goggles must be worn at all times.
- Nitrile gloves must be worn when handling chemicals and when handling chromatography paper.

PRE-LAB QUESTIONS

Name: _______________________________________   Partners: _______________________________________

TA: _______________________________________   Section: _______________   Date: _______________

Answer the following questions using information obtained from lecture, the textbook, or lab manual. The responses will be checked at the start of your lab section for credit. Failure to complete may result in exclusion from participating in the lab.

1. Define **chromatography.**

2. Define **mobile phase.** What is the mobile phase in this experiment? Be specific.

3. Define **stationary phase.** What is the stationary phase in this experiment? Be specific.

4. Draw the zwitterionic structures of the five amino acids to be examined in this experiment.

5. What factors are going to determine how far a particular amino acid travels up the chromatography paper? Predict which class of amino acids (non-polar, neutral polar, acidic, or basic) will move furthest and which will move the least.

BACKGROUND

Almost all substances we come into contact with on a daily basis are impure; that is, they are mixtures. Samples from the environment or from manufacturing often require analysis of some components of the mixture. Similarly, compounds synthesized in the chemical laboratory are rarely produced in a pure state. They are almost always produced with impurities including reaction byproducts and leftover reactants. As a result, a major focus of research in chemistry is designing methods of separating, identifying, and quantifying the various components of mixtures.

Many of these separation methods rely on physical differences between the components of a mixture. Undoubtedly, you are already familiar with several means chemists use to effect separations based on physical differences. These techniques include: filtration, where separation may be effected because substances are present in different states (solid versus liquid); centrifugation, where separation is effected by differences in density; and distillation, where separation is effected by taking advantage of differences in vapor pressure of the various components. In this laboratory exercise, we will examine the separation of amino acids using paper chromatography. Chromatography in general is a highly useful method for separating mixtures and identifying substances. As the name indicates, the chromatographic technique was first developed with colored substances; however, the principles underlying separation of colored substances apply equally well to colorless substances although the separation cannot be followed visually.

Chromatography is a method of separating a mixture of components, or the analytes, based on their relative interaction between two phases. All chromatography techniques have three important components: the mixture of species being separated (**analytes**), a **mobile phase,** and a **stationary phase** (Figure 8.1). The mobile phase is a flowing liquid or gas used to push the analytes over or through a stationary porous material (the stationary phase). Because of physical interactions, such as intermolecular forces, between the analytes and the stationary phase, the analytes move across the stationary phase more slowly than in just the mobile phase. Because physical interactions

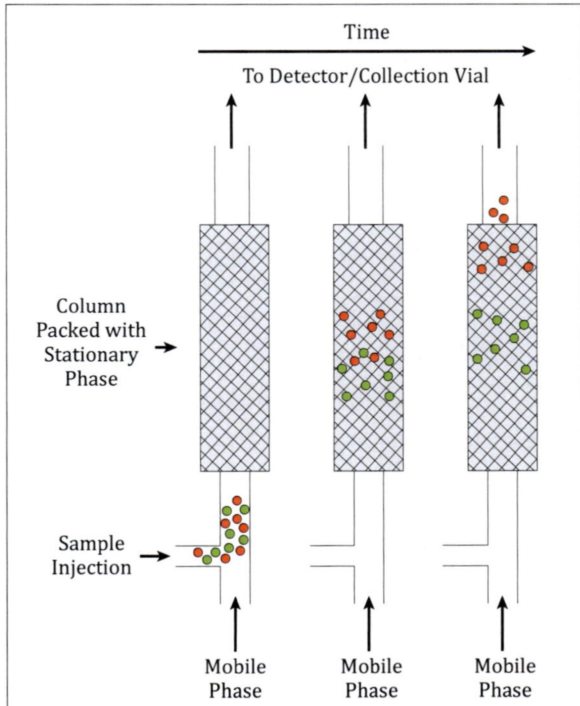

FIGURE 8.1 Liquid Column Chromatography A two component mixture is injected into the flowing mobile phase and swept into a chromatography column containing the stationary phase. Because the ● molecules interact more strongly with the stationary phase than the ● component, they lag behind. As more mobile phase is pumped through the column, the two components are eventually separated and removed from the column.

between the various analytes and the stationary phase can be different for each component, the components move along the stationary phase at different speeds. Those that strongly interact with the stationary phase lag behind those that interact more weakly. As a result, the components of the mixture may be separated.

In paper chromatography, the mobile phase is a liquid, and the stationary phase is the sheet of paper carrying a thin layer of adsorbed mobile phase. A solution of the mixture to be separated or the unknown to be identified is applied to the paper as a small spot near the lower edge of the paper, and the paper is dipped into a shallow layer of the mobile phase. By capillary attraction the mobile phase moves up the paper and flows past the spot. This flow process is called development (or "running") of the chromatogram. The molecules to be separated move repeatedly back and forth between the mobile phase and the stationary phase. Molecules move along at the same speed as the liquid while in the mobile phase and are stopped on the paper while in the stationary phase. A spot may move at the same rate as the solvent front (no affinity for the stationary phase), or the spot may not move at all (no affinity for the mobile phase). With the right mobile phase, however, the spot will travel some intermediate distance. By dividing distance moved by the spot by the distance moved by the solvent front, the **retention factor,** or R_f value, for the spot may be obtained. This is a useful means of comparing an unknown analyte to a compound that is known.

It is also important to consider whether the chromatogram is properly "resolved." A good rule of thumb for judging the resolution of a chromatogram is: if the R_f value is between 0.3–0.7, the spot is "well-resolved"; if the R_f value for a particular spot is far outside this range, its resolution is fairly "poor" and conclusions should only be drawn very carefully. Resolution can also refer to how well-separated the spots are along the paper after the chromatogram has been run. The ideal circumstance is that all the analyte spots are clearly spaced out along the vertical length of the paper. If spots run too similar of a distance, it becomes difficult to draw definitive conclusions, however changing the solvent mixture can often result in better separation.

The paper to be used in this experiment is made up primarily of cellulose which structure is shown in Figure 8.2. From our previous discussion of carbohydrates, cellulose is a polar polysaccharide that contains many —OH and —O— groups that allow for numerous hydrogen bonding and dipole-dipole interactions between the paper and the amino acids. Comparing the structure of cellulose with that of the amino acids, we would expect strong interactions between the amino acids and the stationary phase. Consequently, the mobile phase needs to be quite polar to move the amino acids off the bottom of the paper.

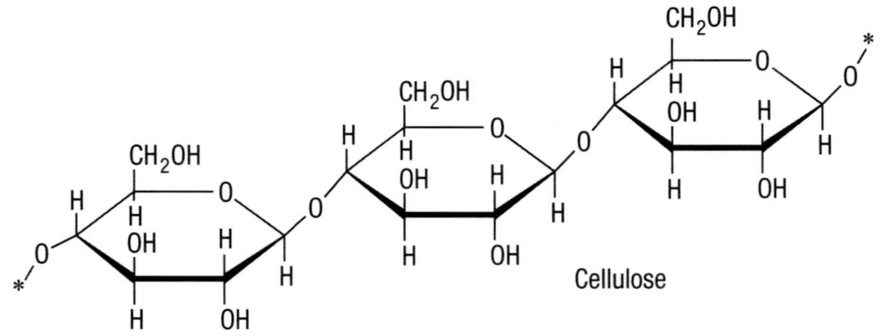

FIGURE 8.2 Portion of the Structure of Cellulose

In this experiment you will examine the chromatographic behaviors of some colorless amino acids. Spots due to the amino acids will be made visible by spraying with a reagent (ninhydrin) which reacts with the amino acids to form a colored product. You will be issued one or more unknowns which will be the same as one of several reference compounds (or perhaps a mixture of reference compounds). These unknowns should be able to be identified with some confidence on the basis of R_f values. Components of amino acid mixtures cannot always be identified with confidence even though the spots are clearly separated because the R_f value of a spot may lie between values for two of the reference compounds. The R_f value of a substance may be increased or decreased due to weak interactions with the other components of the mixture.

PROCEDURE

Paper Chromatography

1. Obtain one 300-mL Berzelius beaker (no spout) or use a 600-mL beaker, aluminum foil to cover the beaker, and one rectangular (10 cm × 20 cm) sheet of filter paper. Gloves should be worn during this procedure to avoid contamination of the filter paper with amino acids from your skin.

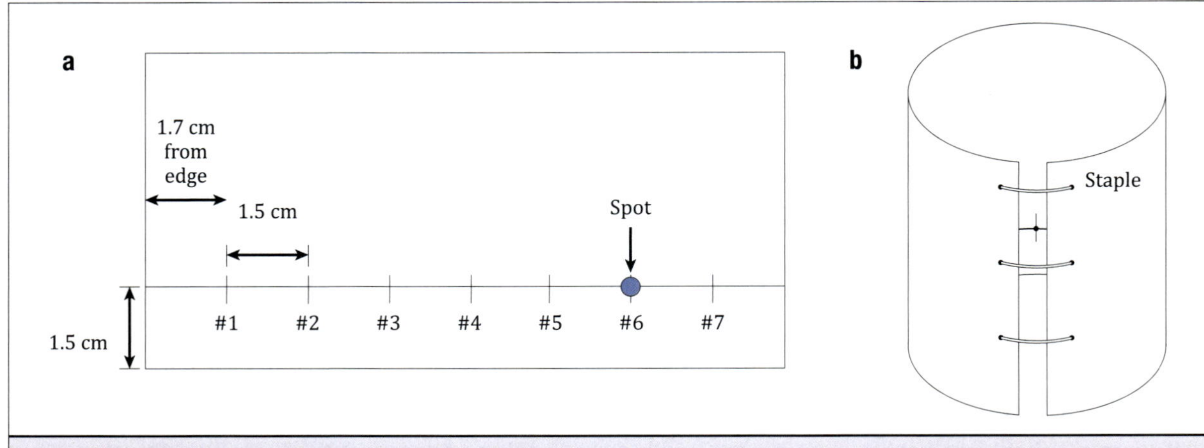

FIGURE 8.3 Paper Chromatogram (a) Schematic layout **(b)** Paper ready for developing chamber.

2. Manipulate the filter paper on a clean sheet of ordinary paper to avoid spurious spots. Use a ruler to draw a faint pencil line parallel to a long side of each piece of filter paper and 1.5 cm from an edge. Beginning about 1.7 cm from either side, make seven small perpendicular marks along each line spaced ~1.5 cm apart (see Figure 8.3).

3. The unknown substances to be chromatographed will be available in small bottles—be sure to return stoppers to their proper bottles when finished. The test substances you should apply to each sheet of filter paper are shown below.

#1. arginine	**#4.** leucine	**#6.** unknown
#2. glycine	**#5.** cystine (dimer of	**#7.** unknown
#3. alanine	two cysteines)	

4. Following the guidance of your TA, use a toothpick or other spotting implement to make small spot of each amino acid solution at each indicated point on your paper. It is advised to practice spotting a few times on scrap filter paper to develop a technique as putting too much sample onto the paper can create messy chromatograms that are hard to interpret.

5. Identify each spot applied by a penciled number under the spot that is keyed to the name of the substance.

6. After the spots have dried a little, form a cylinder from the paper and staple it together as shown in Figure 8.3. Use two staples, about 3 cm from each end of the cylinder. If the sides of the paper are stapled so that they touch, the solvent front will not develop parallel to the base line and the end spots will be skewed.

7. Follow the directions of your TA as to the solvent mixture to use for developing your paper chromatogram. These will either be mixtures of methanol and water or a provided solution that is 60% 1–butanol, 15% glacial acetic acid, and 25% water. Add the appropriate mixture to your beaker making sure that the solvent level will be below the line of spots.

Acetic acid is a highly flammable liquid and vapor that is harmful to skin and eyes.

1–Butanol is a highly flammable liquid and vapor that is harmful to skin and eyes.

8. Place the cylinder of paper spotted with amino acids in this beaker so that the spots are at the bottom and the cylinder does not touch the walls of the beaker.

9. Cover with aluminum foil and allow to proceed undisturbed until the solvent front (which you can see moving) has advanced to within a few millimeters of the top of the paper. This will take a while, perhaps as long as one hour.

10. Once the solvent front has reached the top, remove the cylinder and lightly pencil along the solvent front.

11. Allow a few minutes for the solvent to evaporate from the paper and unstaple the cylinder.

12. Under the guidance of your TA, attach the paper with paper clips to the support provided in a hood. While wearing gloves, carefully spray the paper with the ninhydrin solution. All parts of the paper should be moist with the spray, but never dripping wet.

13. Allow several minutes for the sprayed solvent to evaporate and warm the paper by placing it in the designated oven. Purple spots corresponding to locations of the amino acids should eventually become apparent. Make pencil dots at the estimated centers of each spot.

14. Once visualized, for each of your chromatograms measure the distances from the base line to the spot centers and from the base line to the solvent front. Develop your own methodology for how to consistently judge where/how to make these measurements.

15. Calculate the R_f value for each spot, and record these on your Data Sheet, making sure to note the particular solvent system used to develop the chromatogram.

16. Draw a representation of your chromatogram in your Data Sheet. You might also take a picture of it for future reference or use in your lab report.

17. Share your chromatogram with the other groups in your class. Again with your classmates' chromatograms, calculate R_f values and tabulate the results. Draw representations of these chromatograms as well for use in your report.

DATA SHEET

Name: _________________________________   Partners: _________________________________

TA: _________________________________   Section: _____________   Date: _________________

Chromatogram #1 ————————————————————

Observations and calculations: Mobile-phase: _________________________________

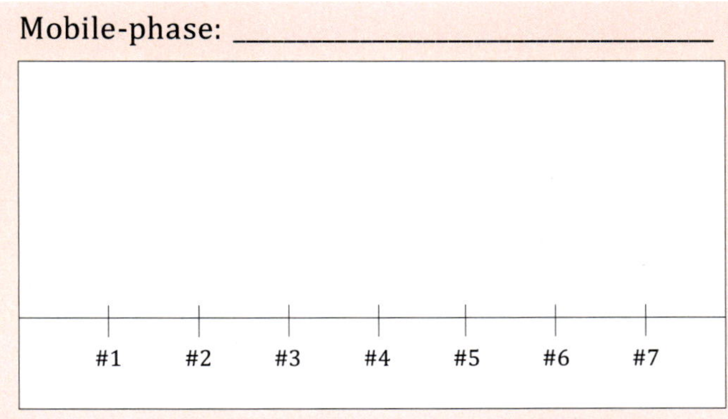

Chromatogram #2 ————————————————————

Observations and calculations: Mobile-phase: _________________________________

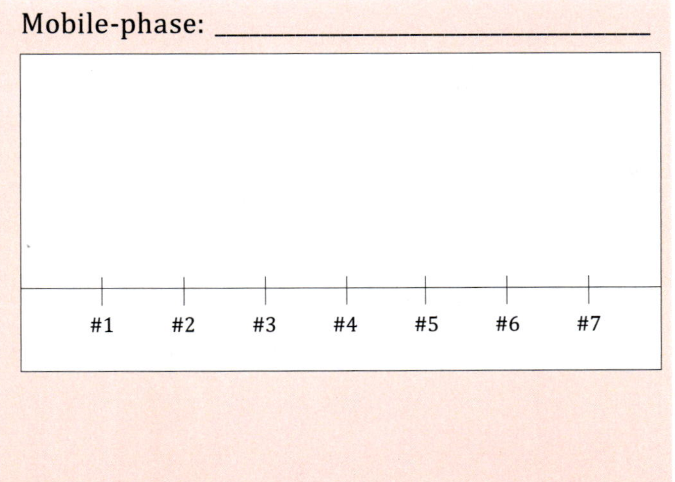

Chromatogram #3

Observations and calculations:

Mobile-phase: _______________________________

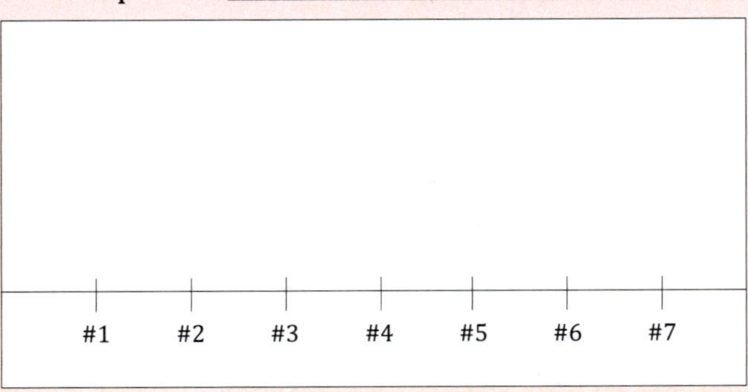

Chromatogram #4

Observations and calculations:

Mobile-phase: _______________________________

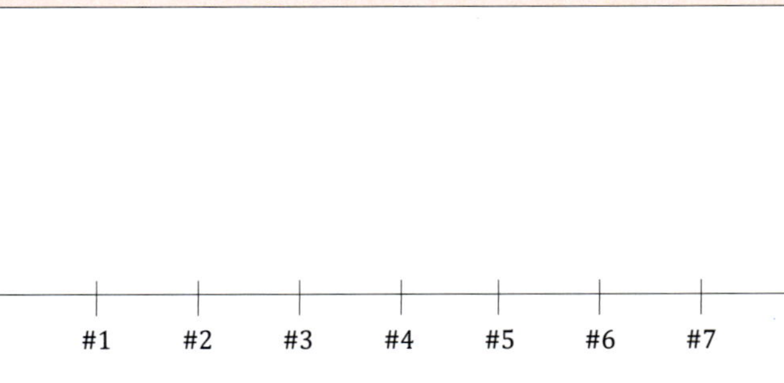

TA Signature	
Pre-Lab Assignment ________ Safety/Participation ________ Lab Write-Up ________	Ask your TA to review your work and sign your report. The TA will sign above once satisfied that the student has performed the entire procedure. The report will not be accepted or graded unless signed.

POST-LAB

The post-lab assignment for this experiment is a lab report. Broad directions are provided in Appendix B of this lab manual. Your TA should have some additional suggestions or guidelines that specifically pertain to this experiment. As they are the ones who will be grading the report, it is important to follow their specific requirements.

What follows here are some ideas to consider as you are writing your reports:

- What effect does changing the polarity of the solvent have on R_f values and how far the spots progressed?

- How does the side-chain of the amino acid effect this distance traveled? Is there a broad trend?

- There can sometimes be difficulty in practically calculating R_f values. What ultimate criteria did you use to judge them?

- What can be said about the resolution of your chromatogram? Were the individual amino acids easy to distinguish via the chromatogram or did they run the same distance? How might you improve the resolution?

- What are the likely identities of your unknown amino acids? How definitive are these conclusions?

- What aspects of the experiment would you change to achieve better results? Why?

Amino Acids and Proteins

Learning Objectives

- Recognize that amino acids are the building blocks for proteins.
- Understand protein hydrolysis and the structural form of its products.
- Connect the physical properties of amino acid and proteins to their chemical structures.
- Formulate conclusions based on qualitative experimental observations.

Additional Reading and References

- Timberlake: 19.1, 19.2, 19.5

Safety Precautions and Hazards

- Chemical splash goggles and lab coat must be worn at all times.
- Nitrile gloves must be worn when handling chemicals.

PRE-LAB QUESTIONS

Name: ________________________________   Partners: ________________________________

TA: ________________________________   Section: ______________   Date: ______________

Answer the following questions using information obtained from lecture, the textbook, or lab manual. The responses will be checked at the start of your lab section for credit. Failure to complete may result in exclusion from participating in the lab.

1. **a.** Define ***amphoteric (or amphiprotic).***

   b. Illustrate the amphoteric characteristics of amino acids by writing the reactions of

   i. alanine with HCl.

   ii. alanine with NaOH.

2. Identify ***all*** the acids and bases in the above reactions.

3. What kinds of amino acids turn yellow upon the addition of nitric acid? Explain why.

4. What is the structural difference between proteins and amino acids that results in a positive result for the biuret test for one and not the other?

5. Casein will be isolated from milk in this experiment by acidifying the protein and then precipitating it in alcohol. Why do you think the acidification part of the procedure is necessary? Explain how selective solubility is used to separate fats and proteins.

BACKGROUND

Proteins are complex, natural polymers which upon hydrolysis yield a mixture of amino acids, some 20 of which are common. In this experiment, a protein will be isolated and a few chemical properties of amino acids and proteins will be illustrated. A generalized scheme below shows the hydrolysis of a protein to form amino acids.

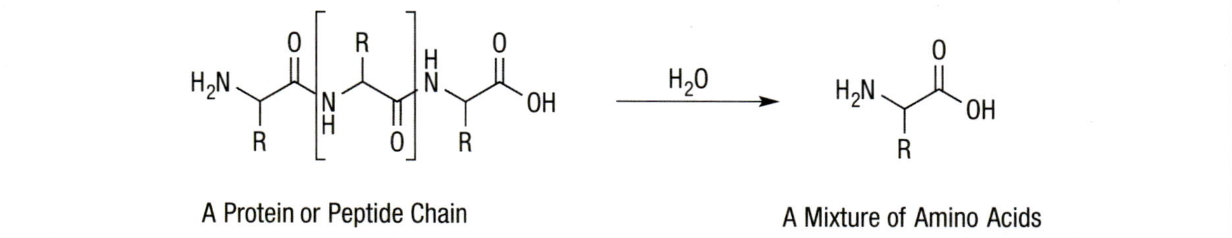

A Protein or Peptide Chain   A Mixture of Amino Acids

FIGURE 9.1 Hydrolysis of Peptides or Proteins R indicates an amino acid side-chain.

The structural form shown above is a poor representation of the structure of amino acids. Due to the acid-base characteristics of the carboxyl and amino groups, amino acids are commonly found in a structural form called a **zwitterion** (Figure 9.2). Many amino acids are soluble in water, mainly because of salt-like properties which result from this dipolar structure.

Amino acid solutions are roughly neutral, unless the R— group contains another amino or carboxyl group, in which case the solution will be alkaline or acidic, respectively. In some cases, particularly when R— has many carbon atoms or is aromatic, the amino acid may be only slightly soluble in water.

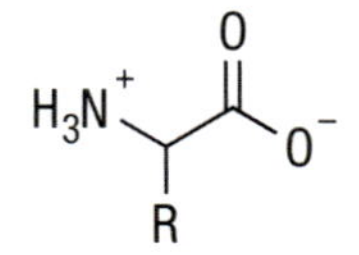

FIGURE 9.2 Zwitterion Structure of Amino Acids

Amino acids can function as acids or bases, forming salts with either strong bases or strong acids:

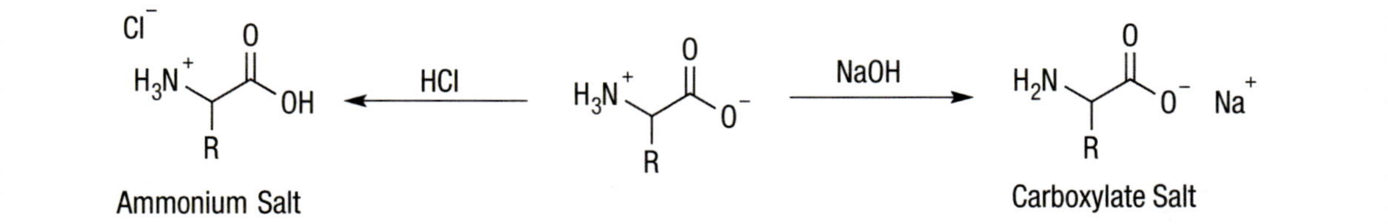

Ammonium Salt   Carboxylate Salt

FIGURE 9.3 Amphoteric Nature of Protein Chains

Such substances that can act as either acids or bases are described as **amphoteric.** Protein chains may also be amphoteric despite the fact that the amino and carboxyl group of each of the amino acids in the middle of the chain have been reacted to form the peptide backbone. This is because some amino acids have two carboxyl or amino groups; these are part of the R— groups attached to the main chain and are free to react with added base or acid.

In this particular experiment, characteristics of amino acids and proteins will be explored using a protein called **casein;** the principal protein found in milk. Casein may be isolated from milk upon acidification which causes the precipitation of casein along with some fat. Since the fat is soluble in alcohol while the casein is not, the fat may be removed by washing the precipitate with alcohol to yield a solution containing the milk protein.

There are number of different qualitative tests that may be used to probe the structural characteristics of proteins like casein as well as simple amino acids. Some tests such as observing the ability of amino acids and proteins to bind heavy metals like mercury (Hg^{2+}) and lead (Pb^{2+}) are beyond the scope of this course, but merit brief discussion. Due to the lone pairs of electrons found on the nitrogens of the amine groups, amino acids and proteins form fairly strong coordinate bonds with heavy metal ions. This coordination event often results in the precipitation of the newly formed complex.

Another test that is more useful in distinguishing proteins from amino acids is their reaction with nitrous acid. As shown in the reaction below, nitrous acid reacts with the free amine groups present in either amino acids or proteins to produce nitrogen gas. The amount of nitrogen gas evolved is based on the number of free amine groups in the molecule. Consequently, if the reaction of a protein with nitrous acid were to be compared to that of a hydrolyzed protein, a much greater volume would be noted for the hydrolyzed protein due to the presence of more free amine groups.

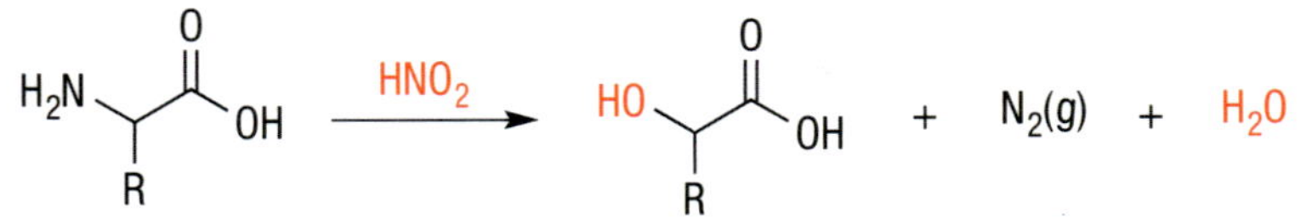

FIGURE 9.4 Reaction of Proteins or Amino Acids with Nitrous Acid

Two separate tests will be focus of your studies for this laboratory experiment. The first of these is the **xanthoproteic test,** which tests for the presence of aromatic amino acids like tyrosine or phenylalanine. These are amino acids present in most proteins. When a protein is treated with concentrated nitric acid, the aromatic R— groups may be nitrated to yield yellow or even darker colored nitro compounds depending upon the solution basicity. The xanthoproteic reaction is very useful in determining whether proteins are present in a given solution. In this experiment, a variety of fabric materials will be tested to identify whether they are protein-based or not.

The other test to be explored is the **biuret test,** which serves to detect the presence of peptide bonds. As with some of the other quantitative test explored in previous experiments, the biuret test involves the complexation of metal ions with a particular reactant to form a colored product. The biuret reagent contains copper ions that, in alkaline solution, form reddish-violet complex ions with biuret or other compounds containing similar structural features. As shown below, the eponymous biuret can be generated from heating urea:

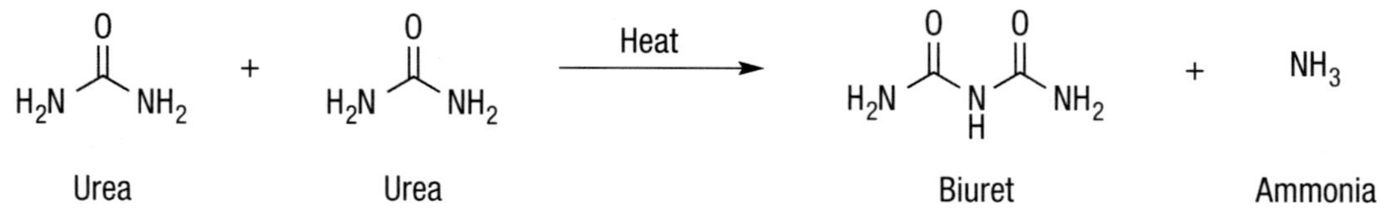

FIGURE 9.5 Conversion of Urea to Biuret

Proteins contain an arrangement of atoms in the form of peptide bonds that are similar enough to biuret such they will give a positive test. Amino acids, however, do not. In this experiment, you will look a number of materials and explore the structural characteristics required for a positive result for the biuret test.

PROCEDURE

A. Isolation of Casein from Milk

1. Accurately weigh approximately 10 g powdered milk. Record the actual amount used on your Data Sheet.

2. Mix the powdered milk with 40 mL of water in a 125-mL Erlenmeyer flask.

3. Heat the solution in a water bath until it reaches 40°C.

4. While stirring, add 35 to 40 drops glacial acetic acid dropwise until the casein forms as a precipitate. Check with your TA prior to proceeding to the filtration.

Acetic acid is a flammable liquid and vapor that causes severe skin burns and eye damage.

5. After at least 5 minutes, filter the precipitate through cheesecloth using a suction filtration set-up, then squeeze gently to remove most of the remaining water.

6. Add the collected mixture of casein and fat to about 25 mL of 95% ethyl alcohol in a 50-mL beaker.

Ethyl alcohol is highly flammable.

7. Stir the mixture with a glass stir rod to ensure proper mixing.

8. Filter the remaining precipitate using suction filtration and a little ethyl alcohol to rinse the beaker.

9. Once most of the liquid has been removed by the vacuum, place your product on a tared watch glass.

10. Weigh your product and record the amount on your Data Sheet.

11. Set the casein aside for future tests.

B. Xanthoproteic Test

12. Obtain four separate watch glasses. Place a small sample of your dried casein onto the first watch glass. To each of the respective remaining watch glasses place a piece of the following fabrics: wool, nylon, and silk.

13. On each material to be tested, add 1–2 drops of concentrated nitric acid and record any observations.

Nitric acid is an oxidizer that causes severe burns and eye damage.

C. The Biuret Test

For this part of the procedure, five substances will be tested using the biuret test: melted urea, urea, reconstituted milk (casein protein), glycine (an amino acid), and hydrolyzed casein. As you proceed, make note of all observations as you will be tasked with comparing the results in the post-lab write-up.

14. Place 0.5 g urea in a dry test tube.

15. Using a test tube holder, heat the tube gently over a low flame until the urea melts and a gas evolves.

16. Note the odor of the gas and test it with a piece of moist pH paper held over the mouth of the tube.

17. Continue gentle heating until gas evolution has almost stopped, and the material shows a tendency to solidify.

18. Allow the tube to cool to room temperature and wash the white solid with 3 mL hot distilled water.

19. Decant the solution, saving the liquid.

20. To the decanted liquid containing the **melted urea,** add 2 mL of 10% sodium hydroxide solution, followed by 2–3 drops of 2% copper sulfate solution.

Sodium hydroxide causes severe skin burns and eye damage.

21. Carefully shake the solution and observe the color.

22. As a control comparison, in a new test tube dissolve 0.5 g **urea** in 3 mL of distilled water and add 2 mL of 10% sodium hydroxide solution followed by 2–3 drops of 2% copper sulfate solution. Compare the result with that observed above. (**Note:** The copper sulfate solution itself has a light blue color, so the mere dilution of this color when this reagent is added to a solution should not be construed as a positive test.)

23. To 1 mL of **reconstituted milk** in a test tube, add 1 mL of 10% sodium hydroxide solution and add 2–3 drops of copper sulfate solution until a purplish violet color appears.

24. To 1 mL of a **3% glycine solution** in a test tube, add 1 mL of 10% sodium hydroxide solution and add 2–3 drops of copper sulfate solution.

Sodium hydroxide causes severe skin burns and eye damage.

25. In an ice bath, cool 1 mL of the provided **hydrolyzed casein.** Add 3 mL of 10% sodium hydroxide solution and several drops of the 2% copper sulfate solution. Compare your results with those of the previous substances.

Hydrolyzed casein contains hydrochloric acid that is corrosive and causes severe skin burns and eye damage.

DATA SHEET

Name: _________________________________   Partners: _________________________________

TA: _________________________________   Section: _____________   Date: _____________

A. Isolation of Casein

Weight of powdered milk: ____________   Percent casein in powdered milk: ____________

Weight of dried casein: ____________

Observations and calculations:

B. Xanthoproteic Test

Substance	Observations
Dried Casein	
Wool	
Nylon	
Silk	

C. Biuret Test (CuSO$_4$)

Substance	Observations
Melted Urea	
Urea	
Reconstituted Milk (casein protein)	
Glycine	
Hydrolyzed Casein	

TA Signature

Pre-Lab Assignment ________

Safety/Participation ________

Lab Write-Up ________

Ask your TA to review your work and sign your report. The TA will sign above once satisfied that the student has performed the entire procedure. The report will not be accepted or graded unless signed.

POST-LAB

1. Based on your collected data, what is the percentage of protein found in non-fat powdered milk? How does this compare to the typical average found in non-fat powdered milk? What might account for any differences?

2. Based on the xanthoproteic test, which of the substances tested contain aromatic R—groups? Explain in detail your reasoning using data collected.

3. What are your conclusions concerning the acid hydrolysis of a casein and its effectiveness?

4. Write an equation for the reaction of glycine with nitrous acid.

5. Assume that an experiment was performed where a sample of casein protein and a sample of hydrolyzed casein were subjected to a reaction with nitrous acid. Predict the results you would see and explain your reasoning.

6. Considering the discussion of solubilities in the introduction and your understanding of amino acids, explain why casein precipitates in acidic solution.

Chromatographic Exploration of Enzyme Activity with Lactase

Learning Objectives

- Explore the activity of lactase at pH levels encountered in the body to determine where it is most active.
- Use chromatography to track the progress of a reaction.
- Recognize the role environmental conditions play on enzyme activity.
- Connect the concepts of carbohydrates, enzymes, and chromatography.

Additional Reading and References

- Timberlake: 20.1, 20.2, 20.3
- Reprinted with permission from Pope, Sandi R.; Tolleson, Tonya D.; Williams, R. Jill; Underhill, Russell D.; Deal, S. Todd. "Working with Enzymes—Where Is Lactose Digested? An Enzyme Assay for Nutritional Biochemistry Laboratories." *J. Chem. Educ.* **1998,** *75,* 761. Copyright 2016 American Chemical Society.

Safety Precautions and Hazards

- Chemical splash goggles and lab coat must be worn at all times.
- Nitrile gloves must be worn when handling chemicals.

PRE-LAB QUESTIONS

Name: _______________________________________   Partners: ____

TA: ____   Section: ________________   Date: __________________

Answer the following questions using information obtained from lecture, the textbook, or lab manual. The responses will be checked at the start of your lab section for credit. Failure to complete may result in exclusion from participating in the lab.

1. Draw the chemical reaction for the hydrolysis of lactose.

2. Why does pH have an effect on enzymatic activity?

3. What is going to be the chromatographic observation that will allow you to determine which pH (and consequently which part of the body) the enzyme lactase is located?

4. Do you expect lactose or glucose to travel farthest up the chromatography paper? Explain your answer.

BACKGROUND

Lactase is the enzyme responsible for digestion of milk sugar in our bodies. It is a hydrolytic enzyme which breaks down the disaccharide lactose into its monosaccharide components—galactose and glucose. The lack of this enzyme in the digestive tract leads to a disorder known as lactose intolerance, a medical condition characterized by the onset of abdominal cramps and diarrhea following ingestion of foodstuffs containing lactose. This condition is virtually nonexistent among infants, but for reasons not entirely clear to medical science, becomes more prevalent after childhood. People who suffer from this condition must refrain from the consumption of dairy products or take dietary supplements of the missing enzyme in order to digest the lactose present in these foods.

In this experiment, you will use a commercially available source of lactase to study the mode of action, as well as the site of physiological activity, of this enzyme. In order to follow the hydrolytic cleavage of lactose to its component monosaccharides, you will use a chromatographic method based on the oxidation of the glucose product. In turn, you will attempt to discover the site of physiological activity by varying the pH of the enzyme solution and monitoring the activity of the enzyme.

The digestive enzymes of the human body are distributed throughout the digestive tract. A few, like amylase, are located in the mouth, a region of neutral pH; some are located in the stomach, a region of acidic pH, ~1; some are located in the first few inches of the small intestine, a region of near neutral pH, ~6–7; and some are located further along in the intestines, a region of slightly basic pH, ~8. By varying the pH of your test solutions to include a range of values you should be able to determine if lactase is located in the stomach, duodenum (beginning of small intestine), or further along in the intestines.

The progress of the reaction of lactose as it is being hydrolyzed can be tracked by spotting a sample of the reaction mixture onto chromatographic paper at various time intervals. Once the chromatogram has been run and developed, the composition of the reaction mixture at each time point may be discerned. To provide more clarity to the results, it useful to also include a spot for the starting material (lactose) and one of the product (glucose). As you are interpreting the results of this experiment, keep in mind those principles of chromatography that were discussed in "Paper Chromatography."

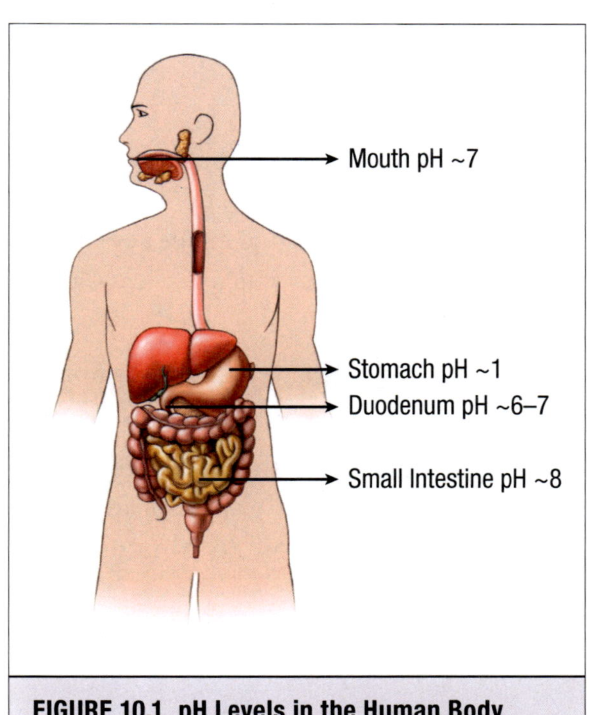

FIGURE 10.1 pH Levels in the Human Body

PROCEDURE

A. Preparation of the Lactase Solution

1. Obtain one lactase pill from your TA and grind it up using a mortar and pestle.

2. In a small beaker, combine the crushed pill and 10 mL of distilled water.

3. Swirl the solution for ~5 minutes, then decant the liquid into a clean beaker and discard the remaining solid.

4. Using pH paper, determine the pH of this solution. Label this solution **Lactase Enzyme Prep.**

B. Preparation of Reaction Solutions

Neutral: Obtain 5 mL of 5% lactose solution and mix it with 2.5 mL of your lactase enzyme prep in a small beaker. Record the time and pH of the solution.

Acidic: Combine 2.5 mL of your lactase enzyme prep with 2.5 mL of 0.1 M HCl. Measure the pH of this solution. Mix 2.5 mL of the resulting solution with 5 mL of 5% lactose in a small beaker. Record the time.

Basic: Combine 2.5 mL of your lactase enzyme prep with 2.5 mL of the buffer solution. Measure the pH of this solution. Mix 2.5 mL of the resulting solution with 5 mL of 5% lactose in a small beaker. Record the time.

C. Preparation of Chromatography Sheet

5. Obtain a paper chromatography sheet from your instructor and lightly (in pencil) label the top corner as "Neutral," "Acidic," or "Basic" based on which of these you were assigned to do.

6. Draw a line across the bottom of the paper chromatography sheet about 1 cm from the bottom of the sheet (as demonstrated by your TA). Making sure to stay approximately 1.5 cm from each edge, place seven "x"s equally spaced across the line and label them "L" for lactose, "G" for glucose, "L + G" for a co-spot of lactose and glucose, "10," "20," "30," and "40."

7. Using the spotting technique demonstrated by your TA, spot the "x" labeled "L" with the 5% lactose solution and the spot labeled "G" with the 5% glucose solution.

8. At the appropriate time, remove a small amount of the solution from your reaction solution and spot it on the appropriate paper chromatography sheet. (One spot is placed after 10 minutes, one after 20 minutes, etc.)

9. Continue until all "x"s have been spotted at the appropriate times.

D. Development of the Chromatogram

10. After the final spot has dried, form a cylinder from the paper and staple it together as shown. Be sure to leave a space between the edges of the sheet so they do not overlap.

11. Add the eluting solution (3:1 mixture of acetonitrile and 0.1 M ammonium acetate) to the bottom of a large (600-mL) beaker. Make sure that the level of the solution is below the line of spots.

> **The eluting solution is a flammable mixture and its vapors should not be inhaled.**

12. Place the cylinder of paper spotted with the sugars in this beaker so that the spots are at the bottom and the cylinder does not touch the walls of the beaker.

13. Cover the beaker and allow to stand undisturbed until the solution has advanced to within a few millimeters of the top of the sheet.

14. Remove the chromatogram from the developing chamber and allow to dry for about 10 minutes. While your sheet is drying, you should prepare your first visualization solution as follows:

Visualization Solution #1: Prepared by combining 1.0 mL of saturated silver nitrate with 19 mL of acetone in a tray. A white solid (silver nitrate) will form in the bottom of the tray. Water should be added dropwise, with careful stirring until the white solid just dissolves.

> **Silver nitrate is an irritant and toxic to aquatic life. Acetone is highly flammable and its vapors should not be inhaled.**

15. When your paper chromatography sheet is dry, grasp it with a pair of tongs and carefully dip it into the first visualization solution so that the entire paper gets wet, and allow to dry again (~10 minutes).

16. Once the sheet has dried, dip the sheet (using tongs) into the solution labeled **Visualization Solution #2.**

Visualization Solution #2: A sodium hydroxide solution in ethanol, prepared by dissolving sodium hydroxide in water and diluting with 95% ethanol.

> **Ethanol is highly flammable and sodium hydroxide causes severe skin burns and eye damage.**

17. Taking care not to touch the paper chromatography sheet with your hands, place it on several layers of paper towels to dry.

18. Draw the resulting chromatogram on the Data Sheet and share the results with the other groups in lab.

DATA SHEET

Name: ________________________________ Partners: _________________________________

TA: __________________________________ Section: ______________ Date: ________________

Draw pictures of the three chromatograms in the area below.

Neutral

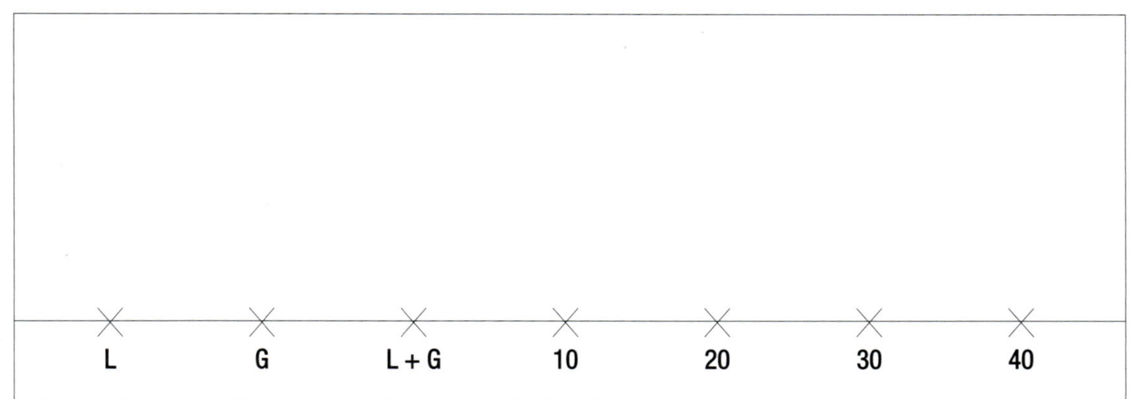

Acidic

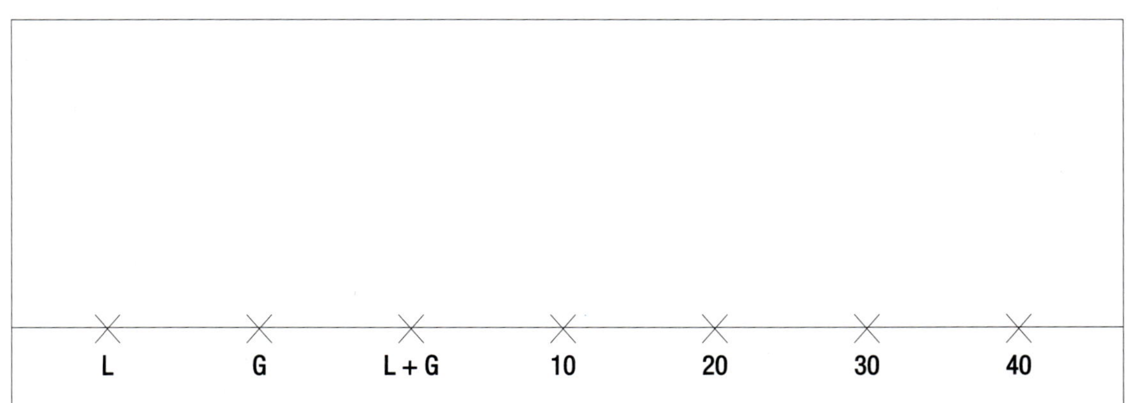

Basic

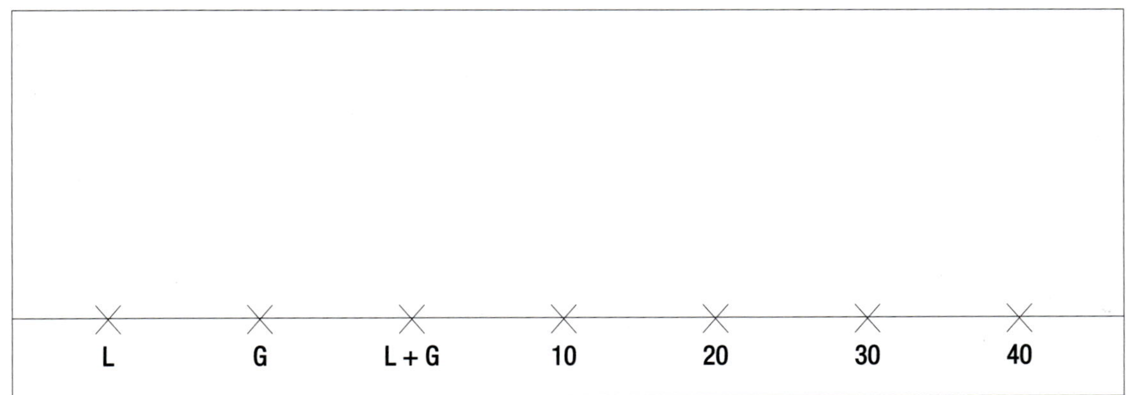

Notes and Observations

<table>
<tr><td colspan="2">TA Signature</td></tr>
<tr>
<td>

Pre-Lab Assignment ________

Safety/Participation ________

Lab Write-Up ________

</td>
<td>

Ask your TA to review your work and sign your report. The TA will sign above once satisfied that the student has performed the entire procedure. The report will not be accepted or graded unless signed.

</td>
</tr>
</table>

POST-LAB

1. Which reaction is catalyzed by lactase? Write out the reaction following the general form:
E + S ⇌ ES complex → EP complex → E + P

2. Explain why glucose was included on the chromatogram when lactase acts on lactose.

3. Use the results to estimate the optimum pH of lactase. Explain your reasoning for full credit.

4. Using your three drawings, determine where in the body the lactase enzyme is the most active. Explain your reasoning for full credit.

5. Using your three drawings, determine how quickly lactase begins to work in the body. Explain your reasoning for full credit.

Extraction and Identification of DNA

Learning Objectives

- Understand and perform an extraction using strawberries.
- Confirm the presence of DNA in an extracted sample by testing for the presence of deoxyribose.
- Write a well-constructed lab report documenting the experiment.
- Have some fun enjoying science!

Additional Reading and References

- Timberlake: 21.1, 21.2, 21.3
- Modified from: *A Laboratory Manual for General, Organic, & Biological Chemistry* 7th Edition by Henrickson, Byrd and Hunter (2011).

Safety Precautions and Hazards

- Chemical splash goggles and lab coat must be worn at all times.
- Nitrile gloves must be worn when dispensing Dische's reagent.

PRE-LAB QUESTIONS

Name: _________________________________   Partners: _________________________________

TA: _________________________________   Section: _____________   Date: _____________

Answer the following questions using information obtained from lecture, the textbook, or lab manual. The responses will be checked at the start of your lab section for credit. Failure to complete may result in exclusion from participating in the lab.

1. Describe what a positive test using Dische's reagent is indicative of and what will be observed.

2. Why are strawberries a particularly useful source of DNA for extraction purposes?

3. Why is it important to make the extraction solution cold?

4. Upon extraction, describe what the DNA of the strawberries is going to look like and where it will be located.

BACKGROUND

In 1868, Johann Miescher extracted substances from the nuclei of the cells of humans and salmon. Because these substances were mildly acidic, he called them nucleic acids. As time passed, and with further study, the substances came to be known in the early 1950s as deoxyribonucleic acids (DNA) because of the products they produced when hydrolyzed (reacted with water) in strongly acidic solution. No matter what the source of DNA, the products were always the same (Figure 11.1):

- deoxyribose, a pentose sugar

- phosphoric acid, H_3PO_4

- a collection of four nitrogen-containing bases (Figure 11.2)

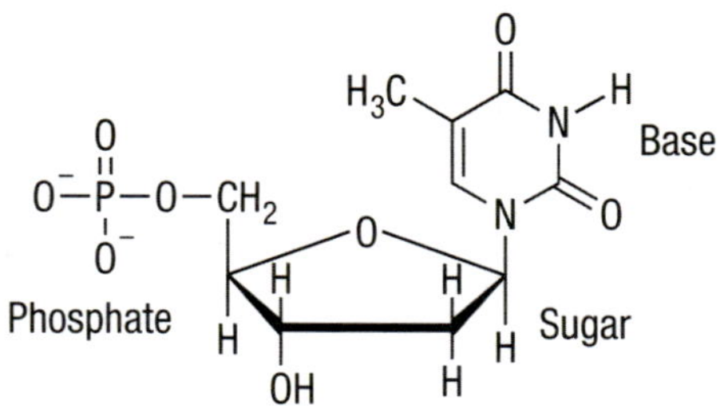

FIGURE 11.1   Representative Nucleotide Structure As shown: deoxythymidine monophosphate.

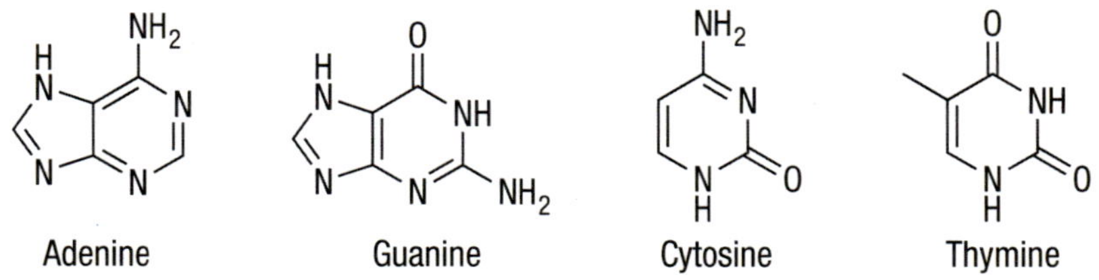

FIGURE 11.2 Nitrogen-Containing Bases Found in DNA

In order to study DNA, it must first be isolated from cells, and that is what you will be doing in this experiment. The DNA extraction solution to be used is just a simple mixture of water, soap, and salt. The solution breaks down the cell and nuclear membranes, freeing the DNA-containing materials.

Though DNA is present in all living things, strawberries provide an unusually rich source, since each cell of a strawberry contains *eight* copies of each chromosome (a DNA-protein structure). Only a few strawberries are necessary to provide a good supply of DNA. Once the DNA is isolated, a chemical test will be done to detect the presence of deoxyribose, verifying that you isolated DNA.

A unique component of the DNA molecule is the pentose sugar, deoxyribose, which, along with phosphate, forms the backbone of the DNA strand. The presence of deoxyribose can be demonstrated with a simple color test using Dische's reagent, an acidic solution of diphenylamine. The appearance of a green to blue or purple color indicates the presence of deoxyribose. A sample of commercially available deoxyribose will also be tested as a comparison.

PROCEDURE

A. Extraction of DNA from Strawberries

1. Set up a boiling water bath in a 100-mL beaker for use in Part B while you proceed with Part A.

2. Prepare an ice bath by filling a 600-mL beaker with ice, then fill it about two-thirds full of tap water.

3. Pour about 20 mL of ethanol into a medium-sized test tube (16 × 150 mm) and set it in the ice bath to chill.

4. In a second medium-sized test tube obtain about 10 mL of the DNA extraction solution. Also set this in the ice bath to chill.

5. Gather six layers of cheesecloth cut into a square and make sure it fits into a large funnel.

6. Attach a large test tube (25 × 200 mm) to a ring stand and set the funnel and cheesecloth into the top of the large test tube.

7. Place three freshly washed strawberries in a clean plastic bag with a zip closure. Remove as much air as possible from the bag and seal it shut.

8. For 2 minutes, mash the strawberries until they become a fine paste.

9. Add the 10 mL of cold DNA extraction solution you placed in the ice bath earlier to the bag containing the mashed strawberries.

10. Knead the strawberry-extraction solution mixture in your hands for at least 2 minutes being careful not to break open the seal on the bag.

11. After 2 minutes, carefully transfer the contents of the plastic bag into the funnel, collecting the pink filtrate in the large test tube below. You will likely want to help force it through the cheese-cloth to speed up this process.

12. Once the liquid stops flowing through the cheesecloth remove the test tube from the ring stand and immediately place it into the ice bath. This will help stop the breakdown of the DNA.

13. After about 3 minutes in the ice bath, remove the test tube.

14. While holding it at a 45° angle, slowly and carefully pour the cold ethanol down the side of the test tube *so that it forms an alcohol layer on top of the pink extraction solution.* Do not agitate the mixture as you return it to the ice bath. Keep the sample cold.

15. While the DNA is cooling, place an additional 5 mL of ethanol into a medium test tube and set it in the ice bath.

Ethanol is highly flammable.

16. At this point you will be able to clearly see the interface where the pink layer on the bottom contact the colorless alcohol layer. In a short time, you should start to see a white mucus-like material gather at this interface. This is the DNA from the strawberries.

17. Using a wire loop, remove the mass of DNA from the interface and transfer it into the test tube containing 5 mL of ethanol. Keep this test tube of DNA in the ice bath for use in Part B.

B. Testing for the Presence of Deoxyribose

18. Place half a pipette of 2.0 M sulfuric acid in a small test tube. Label as **Sample 1.**

Sulfuric acid is corrosive and causes severe skin burns and eye damage.

19. Place half a pipette of 2.0 M sulfuric acid in a second small test tube. Remove about half the DNA you isolated in Part A and place in the acid solution. Gently mix the solution, label as **Sample 2.**

20. Place half a pipette of 2.0 M sulfuric acid in a third small test tube. Add approximately 0.1 g of powdered deoxyribose, label as **Sample 3.**

21. Put all three test tubes into the *boiling* water bath. Keep them there for 15 minutes, then remove from heat and allow to cool in a test tube rack for ~5 minutes.

22. After test tubes have cooled slightly add a full pipette of Dische's reagent to each test tube.

Dische's reagent contains diphenylamine, which is an environmental hazard that must be disposed in a waste container. It also contains two concentrated acids that are corrosive and cause severe skin burns and eye damage.

23. Return them to the boiling water bath for an additional 15 minutes. Record your observations and conclusions regarding the presence of DNA or its components.

24. Dispose of all waste in the appropriate waste container.

DATA SHEET

Name: _______________________________   Partners: _______________________________

TA: _______________________________   Section: _____________   Date: _____________

Test Tube	Observations	Conclusions
#1 Sulfuric Acid		
#2 Sulfuric Acid + Extracted DNA		
#3 Sulfuric Acid + Deoxyribose		

Notes

<table>
<tr><td colspan="2">TA Signature</td></tr>
<tr><td>
Pre-Lab Assignment _________
Safety/Participation _________
Lab Write-Up _________
</td><td>
Ask your TA to review your work and sign your report. The TA will sign above once satisfied that the student has performed the entire procedure. The report will not be accepted or graded unless signed.
</td></tr>
</table>

POST-LAB

The post-lab assignment for this experiment is a lab report. Broad directions are provided in Appendix B of this lab manual. Your TA should have some additional suggestions or guidelines that specifically pertain to this experiment. As they are the ones who will be grading the report, it is important to follow their specific requirements.

What follows here are some ideas or questions you might consider as you are writing your report:

- Why were the strawberries smashed before the DNA extraction solution was added?

- Why was soap a necessary part of the DNA extraction mixture?

- Where was the DNA present in the extraction? The more aqueous layer or ethanol?

- Why was acid added before your extracted DNA could be tested for the presence of deoxyribose?

- What does the experiments in test tube #1 and #3 tell you about the results of #2. Why are these important?

- How might we get more quantitative data out of this experiment? Describe how might this be achieved.

Chemical Reagent Usage Overview
1020 Laboratories

Experiment 1: Molecular Models: Organic Compounds

No chemical reagents used.

Experiment 2: Dyes and Dyeing

- malachite green
- p-nitroaniline
- hydrochloric acid (concentrated)
- Congo Red
- sodium nitrite
- sodium carbonate
- 2–naphthol
- tannic acid
- 5% sodium hydroxide

Experiment 3: Alcohols, Aldehydes, and Ketones

- 0.1 M potassium permanganate
- ethanol
- 2-propanol
- tert-butanol
- benzaldehyde
- acetone
- methanol
- 1-butanol
- 1-pentanol
- cinnamaldehyde
- hexane

Experiment 4: Identification of an Unknown Carbohydrate

- starch solution, aqueous
- Benedict's reagent
    - one liter of Benedict's solution contains 173 grams sodium citrate, 100 grams sodium carbonate, and 17.3 grams cupric sulphate pentahydrate
- iodine solution
    - one liter of iodine solution contains 3 grams iodine and 15 grams potassium iodide
- Barfoed's solution
    - One liter of Barfoed's solution contains 70 grams copper acetate monohydrate and 9 milliliters of glacial acetic acid
- Seliwanoff's reagent
    - One liter of Seliwanoff's reagent contains 1 gram of resorcinol and 330 milliliters of concentrated hydrochloric acid
- glucose
- fructose
- sucrose
- lactose

Experiment 5: Synthesis of Organic Compounds: Aspirin and Wintergreen Oil

- salicylic acid
- methanol
- acetic anhydride
- ethanol
- sulfuric acid

Experiment 6: Fats, Oils, Soaps, and Detergents

- lard
- phenolphthalein solution
- ethanol
- mineral oil
- sodium hydroxide
- 1% calcium chloride
- sodium chloride
- 1% magnesium chloride
- hydrochloric acid
- 1% ferric chloride

Experiment 7: Analysis of Organic Compounds

- universal indicator
  - as reported in its MSDS: 76.00% denatured ethyl alcohol, 23.85% water, .06% bromothymol blue, .06% phenolphthalein, .005% thymol blue, and .025% methyl red
- ethanol
- 0.1 M potassium permanganate
- sodium bicarbonate
- ceric ammonium nitrate reagent
  - contains 200 grams ceric ammonium nitrate in 500 milliliters 2.0 M nitric acid
- 1% ferric chloride solution
- 2,4–dinitrophenylhydrazine reagent
  - contains 3 grams 2,4–dinitrophenylhydrazine, 15 milliliters sulfuric acid, 20 milliliters water, and 70 milliliters ethanol
- Benedict's reagent
  - one liter of Benedict's solution contains 173 grams sodium citrate, 100 grams sodium carbonate, and 17.3 grams cupric sulphate pentahydrate
- 3 M sodium hydroxide
- acetic acid
- acetone
- cotton seed oil
- glucose
- mineral oil
- phenol
- benzenesulfonyl chloride
- isobutylamine

Experiment 8: Paper Chromatography

- glycine
- alanine
- leucine
- 2% ninhydrin
    - contains 2 grams ninhydrin per 100 milliliters acetone
- eluting solution
    - contains 60% 1–butanol, 15% glacial acetic acid, and 25% water
- cystine
- arginine

Experiment 9: Amino Acids and Proteins

- powdered milk
- 1% sodium nitrate
- glacial acetic acid
- urea
- 95% ethanol
- 10% sodium hydroxide
- 1:1 ether-ethanol mixture
- 2% cupric sulfate
- 3% glycine solution
- nitric acid (concentrated)
- glycine
- 10% hydrochloric acid
- 5% sodium nitrite

Experiment 10: Chromatographic Exploration of Enzyme Activity with Lactase

- 5% glucose solution
- 5% lactose solution
- 0.1 M hydrochloric acid
- saturated silver nitrate
- 3:2 acetonitrile and ammonium acetate solution

Experiment 11: Extraction and Identification of DNA

- Dische's reagent
    - contains 1 gram of diphenylamine, 2.5 milliliters sulfuric acid, and enough acetic acid to make 100 milliliters
- 2 M sulfuric acid
- deoxyribose powder

Lab Report Writing Suggestions

One of the most important things about science is communicating results of an experiment with other interested individuals. This is a skill that has broad applications even outside of traditional scientific fields. Throughout this semester, you will be performing experiments that require the collection of data and for observations to be made. Lab reports are the most common method for presenting your findings and discussing what they mean. This semester you will have the opportunity to write four lab reports, presenting your understanding of the performed experiments. For students in CHM 1020, this can be a daunting task as most individuals have not had all that much experience (if any) in scientific writing, let alone in the specific format of a lab report. Each report will be an opportunity to refine your skills and make tangible connections between the laboratory experiments and the larger organic chemistry content being discussed in class. The first couple reports may be a bit rough, but hopefully throughout the process you will be able to refine your skills and produce solid, well-constructed reports by the end of the semester.

The (grading) focus for each report will be slightly different:

- **Dyes and Dying:** general format, writing-style, and breadth of analysis

- **Synthesis of Aspirin:** clarity and fullness of discussion/conclusions, modifying experimental procedures

- **Paper Chromatography:** making connections to lecture content and expounding upon them

- **Extraction and Identification of DNA:** overall report quality

What follows are some dos and don'ts about lab report writing that may be useful as you are thinking about putting together yours. These are organized by section which should correspond to those sections you are including in each report. Your teaching assistants will also discuss with you those specific aspects that are important for each experiment. They are ultimately the ones who will spend the time reading and scoring the reports, so be mindful of their suggestions/requirements as well.

A Perspective on Important Aspects to Lab Report Writing

(Compiled from the experience of years of looking at organic students' lab reports.)

Introduction (~3 points)

a. Does not need to be long, roughly one paragraph long. Anything extensively more than is likely too long and superfluous.

b. It should talk generally about the concepts being covered in the lab experiment and the purpose of the experiment.

c. If any especially relevant lab techniques were utilized, they may also be introduced or described here.

d. This should be written in the present tense and absolutely should not start with the phrase: "The purpose of this experiment is …" or any variation of this.

e. Students often are tempted to lift whole passages from the lab manual. Don't do this. Points will be deducted and academic integrity violations could be instigated if especially grievous.

f. A simple reaction scheme often is included here, if appropriate (like in Experiment 5: Synthesis of Organic Compounds: Aspirin and Wintergreen Oil).

Experimental (~5 points)

a. This section should be written in the past tense and be roughly one paragraph in length with everything written in paragraph form … no bulleted lists.

b. It is not necessary to write in detail about every part of the experimental process. Things that would be common knowledge for the experimenter can be left out. (Example: You wouldn't describe how compounds were measured out.) This is a section you will likely want to have conversations with your TA about prior to turning in the report.

c. This should be written to take into account any changes or deviations that were incorporated into the procedure on lab day. It should not read as a direct copy of the lab book. You may want to reference the lab manual and document only the significant changes that were made … and that is fine if done properly with lab manual properly referenced.

Results and Discussion (~8 points)

a. This should be written in third-person past tense and likely is multiple paragraphs in length.

b. Students should critically discuss their data that might have been collected.

c. The ***Results and Discussion*** seeks describe some basic points ... did the experiment achieve the expected results? What information was learned? This section should be geared toward answering questions such as these (though not explicitly). Some data collected may support or refute a particular finding, but it is important to discuss what each piece of data tells you about your experiment and why you came to that thought-process. Simply giving a bunch of results without explanation does not make for a very good discussion of results. Ideally, the ***Results and Discussion*** should account for about half the report.

d. Depending on the experiment, you may have a lot of data to work with, or very little. Don't worry so much about length. Clarity and cohesiveness is far more important than length.

e. In grading, the TAs are going to be looking at three different aspects:

    i. Presentation of data,

    ii. Demonstrated understanding of the experiment, and

    iii. Logical organization of discussion.

Conclusions (~4 points)

a. This should be approximately one paragraph in length and is usually written in present tense, though it can be done sometimes in the past tense.

b. In the ***Conclusions,*** the overall effectiveness of the experiment is described based on collected results that were just explained above. There were likely aspects that were more successful than others. This is the place to reflect upon the statements made in the ***Introduction*** and whether you approached your stated objectives for the experiments performed.

c. Also, it is advisable to give some ways in which to improve the laboratory if it were to be performed again, perhaps additional data that could have been collected or additional experiments you think might be helpful or interesting. These statements are especially important when things don't go so well with your experiment as it shows you've made some true reflections.

d. Overall, the best ***Conclusions*** sections will tie everything up, referring back to the fundamental principles discussed in the ***Introduction*** by using the results presented in the ***Results and Discussion.***

Citations

You will likely always need to cite information from a source like the lab manual. Guidelines are provided below and should be followed. Points will be deducted for failure to give appropriate references. You should consult with your TA as to their preferred in-text citation format.

Some General Things to Consider

Plagiarism

All reports should be written individually and in each students own words. Lab reports will be submitted electronically to have the originality checked. Penalties will be severe if a particularly egregious lack of originality is found.

Spelling and Grammar

Silly grammar and spelling mistakes really detract from your overall report. It's hard to take results seriously if there are a lot of spelling or grammatical errors. TAs will closely look at these aspects of your reports.

Double-Space Reports!

This makes the job of giving comments/corrections to your reports that much easier.

No First Person Tense in Lab Reports ...

Third person past tense should be used. "I" is just unacceptable in formal scientific writing. Generally speaking, the introduction should be in the present tense, the *Experimental* and *Results & Discussion* in the past tense, and the *Conclusions* perhaps either one depending on how you do it. (A quick way to lose lots of points is to use "I"s everywhere)

Write Objectively About the Experiment Performed

Make sure you have not added personal qualifications to your analysis. Statements like "the experiment went pretty well," and such, are not appropriate for scientific writing. Also, you should not write in a particularly negative fashion (i.e., whining), complaining about this or that in your conclusions or elsewhere in the lab report. Constructive means to improve the experiment are welcomed, but lack of resources or materials should not be used as reasons for why a particular experiment might have failed.

There should be a total of 20 points for each Lab Report, plus an additional 5 points for the Pre-Lab worksheet, giving a total of 25 points.

Citation Style

The ACS style is a standard method of citation in academic publications that originated with the American Chemical Society (ACS). The printed versions of the ACS style manual are entitled *ACS Style Guide: Effective Communication of Scientific Information*, 3rd ed. (2006), edited by Anne M. Coghill and Lorrin R. Garson, and *ACS Style Guide: A Manual for Authors and Editors* (1997). The guidelines may be found easily online. Shown below are some of the commonly needed formats:

Edited Book (like the lab manual)

Last Name, First Initial.; Last Name, First Initial. Title of Chapter/Experiment. In *Title of Book or Manual;* Last Name, First Initial, Ed; Publishing Company: City, Year; Volume; Pages.

Example of a Lab Manual Citation:

Alum from Aluminum Cans. In *Laboratory Guide for Chemistry.* Grossie, D.A. and Underwood, K., Eds.; Hayden-McNeil: Plymouth, MI, 2014; 7th edition; pp. 17–19.

Article Published in a Journal

Last Name, First Initial.; Last Name, First Initial. *Journal.* **Year,** *Volume,* Pages.

Example of a Journal Citation:

Deno, N. C.; Richey, H. G.; Liu, J. S.; Lincoln, D. N.; Turner, J. O. *J. Am. Chem. Soc.* **1965,** *87,* 4533–4538.

URL (web page)

Author, if available. Title of page as listed on the site. Address of page (date accessed).

Example of Web Page:

SDBS: IR (Liquid Film), benzene. http://riodb01.ibase.aist.go.jp/sdbs/cgi-bin/direct_frame_top.cgi (accessed Apr 2008).

Web Resources:

- *http://en.wikipedia.org/wiki/ACS_style*
- *http://pubs.acs.org/books/references.shtml*
- *http://library.williams.edu/citing/styles/acs.php*
- The full ACS guide is available through the WSU library class page or at: *http://pubs.acs.org.ezproxy.libraries.wright.edu/isbn/9780841239999*

Notes

Notes

Notes

Notes

Notes

Notes

Notes

Notes